Under The Gun

Infantryman, Medic, Tattoo Artist: My Year in Iraq

Malcolm Rios

Bloomington, IN

Milton Keynes, UK

AuthorHouse™
1663 Liberty Drive, Suite 200
Bloomington, IN 47403
www.authorhouse.com
Phone: 1-800-839-8640

AuthorHouse™ UK Ltd.
500 Avebury Boulevard
Central Milton Keynes, MK9 2BE
www.authorhouse.co.uk
Phone: 08001974150

First published by AuthorHouse 3/8/2007

ISBN: 1-4259-2869-2 (sc)

Printed in the United States of America
Bloomington, Indiana

This book is printed on acid-free paper.

This book is dedicated to the men and women of our great nation who serve to make the world in which we all live a better place. Thank you for your sacrifices. And to those who have gone so boldly before us, thank you and God speed.

Special dedication to Harold Smelser (MIA), George K. Evans, George K. Evans, Jr., Bill Evans (POW), Clarence Evans (POW), Giles Evans, Jr., John Popham III, John Popham IV, Mike Tetu, and Giles Evans III

Acknowledgements

First and foremost, I thank God for guiding me through this adventure and bringing me home safely. Thanks to my family for their support and sacrifice while I was away. To my mom for the countless hours she helped me with editing. To my friends, thank you for standing behind me with your support, encouragement, mail, and especially your prayers. To those individuals who became my friends over the months that this adventure took place, thank you for sharing your lives with me and allowing me the opportunity to share my life with you. To the men and women of the 278th Regimental Combat Team, thank God it's over!! We did a good job. Let's move forward knowing that. To the families of all personnel who are in theater, God bless you for your sacrifices.

Contents

Lamar Alexander and I at Camp Shelby Farewell Celebration

Ring of Fire

November 15, 2004

We were one week shy of hopping across the big pond. It was definitely going to be an experience that I'd want to share with future generations so they could learn from our mistakes and victories. It's what we'd trained for since boot camp. We were about to become warriors.

Since October 28th and our return from California, almost three weeks have passed. We were sent home for block leave and were now holed up in the barracks at Shelby. Leave was good. I didn't spend a lot of time chasing down people whom I would have liked to see but I did get to see quite a few friends none the less.

The lights were out in the barracks. A lot of the jokers were snoring their asses off. I despised snoring and didn't know how I was going to be able to make it through the next year surrounded by guys who could snore at the drop of a hat. Ear plugs could only do so much and then one just had to learn to sleep through it. I hoped I'd be able to learn how to do that.

I've done twelve tattoos since we've come back from leave. There was a lot of potential for work at Camp Shelby. Baked Tater got a Celtic cross on his back right shoulder. Mohawk sacked up and got his first one tonight. It was a "death before dishonor" skull and turned out really good.

November 21

Things were heating up as we were getting closer to our ship-out date. We were slotted to hit Kuwait first, acclimate, and then head into the deserts of Iraq. We still didn't have a clue as to where we'd actually be. Iron Troop was slotted for FOB Cobra but there were nasty rumors being passed around that The Deuce, our platoon, was getting cut off for use in other operations by other squadrons. It was a bit nerve-racking trying to prepare for something when we didn't know what to expect and no one in our chain of command was giving us any hint of clarity in the situation.

This week we've played Ultimate Frisbee and done a lot of training evolutions like Leadership Reaction Course and language training. We've also tried to play hard outside the military realm, so hard in fact that we've been locked down twice: once for fighting another platoon and the other because our First Sergeant, Death Pants, felt that the Deuce was out of control and had no discipline. He felt that he needed to instill it in us. I didn't have any clue at that time as to how true his assessment was.

Tattoos have been flowing like cheap wine. I made quite a bit of money in the past week. Most of the money I made from inking folks was given in donations. I never charged the amount that I could have because I didn't have any overhead. The work was as good as any professional shop and it was cheaper.

I planned to go to church in town today with Terrible Teddy. The pastor of the church where we were going had graduated from the same high school as TT. Somehow they had managed to link up while we were in Shelby and he had invited TT and anyone else over to their church to worship anytime we cared to. Terrible Teddy, Lug Nuts, Hamper, and I loaded up and went to Temple Baptist for their first service. It was a really nice place. The pastor, his wife, his son, and his daughter-in-law took us to lunch.

We were sitting about twelve hours away from departing for Kuwait. I've talked to a few friends and Mom and Dad. Mom and Dad seemed to be the ones taking it the hardest. I guess I expected that. Mom was crying when I hung up.

I talked to Sneaky Pete and Sarah last night. It was good to know that there were friends like them in the world. I loved them both. They asked me how I was feeling about going to war. I was certain that I would be anxious until the first bullet flew by me or the first IED detonated.

November 24

We left on the 22nd of November for Kuwait. The sky was overcast and a light drizzle fell on the pines and already muddy grounds of Camp Shelby. Third Squadron of the 278th was finally scheduled to catch up with the rest of the Regiment. We loaded our gear in a semi and then loaded ourselves onto the Bluebird buses that always seemed to be a vital part of military and prison transport. After weighing in, we loaded charter buses for our trip to Gulfport for the plane ride across the pond.

We made two stops along the way: one in Gander, Newfoundland and one in Shannon, Ireland. Both were nice places except that we didn't get to see much of either country. We were only allowed to walk the terminals and it was night. There was snow on the ground in Gander.

Terrible Teddy, Lug Nuts, and LT loading up for the flight across the pond

We arrived at Camp Buerhing, Kuwait at about 2000 after what seemed like a twelve hour bus ride from Kuwait City. We cleared our weapons at the front gate, loaded back onto the bus, and were then given the opportunity to eat chow. The chow hall was decked out in Thanksgiving décor and left a lot of us wishing for home, friends, and family. After gorging ourselves on some unexpected good food we headed over to a neighboring tent for our in-brief.

It was midnight and I wasn't too tired, so Black Messiah and I ventured off to try and phone home. We were successful, although all I got to do was leave a message. It was 1500 in Tennessee and Mom hadn't made it back from school. No telling where Dad was. Later, the Black Messiah and I went to the Cyber Zone and tried out their internet services. It left a little to be desired but at least it was available. For five dollars any Joe could stay on the internet for an hour.

I woke this morning to what I thought was bright sunshine, but when I went outside I saw that there was a sand haze that surrounded the entire camp. I hoped that I would still feel the need to see the beauty beyond the haze throughout my time in Iraq.

I also noticed that in passing military women on base, they didn't make direct eye contact. I didn't know if they were so afraid of what was behind the intent of another person's eyes or they just didn't want to make contact with an unknown male. It was a sad statement to me of how far over the edge the issue of sexual harassment had gone.

Santa meets the Black Messiah

November 26

Yesterday was a fiasco. All day long I was wondering whether or not I was going with 2nd Platoon on their first two months of duty here. From what we were told, the Deuce was going to be split up and attached to 1st and 2nd Squadrons. That wasn't too bad other than the fact that with my going as a medic, the Squadron that I would be re-assigned to could add me to their compliment of medics if they had need. Being assimilated into Charlie Med was something that I did not want to happen. Ultimately, Dark Wing determined that he was going to write me in as an 11B (Infantryman) so I could go with his part of the platoon. I'd still do my medic work but in the eyes of 1st Squadron brass, I was simply a ground pounder.

A lot of Joes were getting sick in Kuwait because it was colder than had we expected and our tent didn't have heat in it. Our sleep systems were great but they could only do so much. I felt inadequate at times because most of my training was for trauma situations. When

Joe came to me for his everyday sniffles and gripes, I usually didn't have what he needed simply because there wasn't enough space for everything in my aid bag. I would then refer him to the medics of HHT or the Docs would tell me just to send him up to TMC (Troop Medical Clinic).

This procedure became a pain in the ass for everyone because it made the medics look entirely inadequate. For example, today a young soldier had an asthma attack. He couldn't find his inhaler and I didn't carry any extra inhalers. The only thing I had was an Epi-Pen and I'd only use that for an extreme emergency. I tried getting something from the HHT guys but they were in formation and when I asked Terrible Teddy, he referred me to the TMC.

Last night I did some tattoos, four to be exact. I did them on Crazy, Odom, Nasty, and Holt. Crazy got the Section 8, Odom got a Godsmack-style sun with his child's initials in the middle, Nasty got some random tribal sun that he had picked out about two minutes before he got the tattoo, and Holt got a cover-up of an old cheese tattoo that said, "Crazy" on his back. His was a moon/sun that I came up with on the fly. He really seemed to like it.

I continued to be asked to give tattoos but people did it in such a way as to make me wish I just had a sign up list. It was the nature of the beast. I wanted Joe to know that when the list was up, tattoos were going to be happening. When the list was not up, no tattoos were going to be done.

NOVEMBER 27

I was told by the 3rd Platoon LT that I would remain attached to Iron Troop. I wasn't going to be heading out with 2nd Platoon after all. The rumor was that they were being reassigned to do a two month adventure somewhere in Northern Iraq. Some of the guys would go to Camp Caldwell and some would go to Camp Bernstein but neither would go with a medic. There was an issue about my being the senior medic and leaving the Troop. Some of the guys from the other platoons jokingly said that they wanted me to stay with I Troop because they could continue to get tattoos when we got to Cobra. I wouldn't have minded that but I was being told that I would have to leave the guys whom I had trained with for the last three months.

The squads were drawing closer to one another. They sat around on different cots and spun tales and over-glorified stories of their past escapades. Everyone was a hundred percent bolder here in the safety of our tents where we had yet to feel the fear of combat. Stories of how each man would kill scores of Hajjis and insurgents reverberated each time someone brought up the topic of battle. I questioned Crowfoot this morning in the chow line about how many of the men in his squad had been in a fire fight before. He said four, all former Marines who had been in the Gulf War.

As I was typing, Dark Wing came up and told me that he was still trying to get me onboard with him and his group of guys. The yo-yo effect was starting to wear on me. I wanted the leadership in our unit to make a decision and stick with it. That was all I was asking. I could deal with their decision. It was their indecision that was driving me up the wall. To me, it was a reflection on their lack of leadership. I hoped that things wouldn't always be like this because it would end up getting a lot of men killed who wanted to go home in a year. I prayed that God would help me with my belief in this system.

NOVEMBER 29

How do I describe 2nd Platoon and do them justice? As with most human interaction, there were cliques within this group of 30 guys. There was a group who called themselves "Section 8." Had the Army decided to do drug tests at any time during our deployment, most of them would have popped positive. One of the splinter groups was the "E-4 Mafia." It was exactly what its name implied. These were the louder guys of all the E-4's and when we were in garrison, they were the ones who were always out chasing skirt. In the sand box they were just loud. There was also a group of guys who were our computer geeks. They were led by Clementine who seemed to be the one with the most knowledge of computer systems. Of course, all of us followed Dark Wing. He knew how to motivate and sell people. He also had an uncanny ability to tell hilarious stories. The faces that he made as he was telling them were as humorous as the stories themselves.

Yesterday I went to chapel services twice. The first was the base service. It was a blessing to be in the company of other believers who wanted to worship. Some Private got up to sing a solo. He wasn't that good but I had to give him credit for stepping out and playing his guitar and singing in front of the crowd. We sang Christmas songs for worship. The Chaplain's sermon was about being prepared for the second coming of Christ.

The second service was given at the Squadron. The 3rd Squadron Chaplain delivered the sermon. We received communion between two tents in the desert. That was kind of strange. Never in my life had I imagined that I'd be taking communion in the middle of the Kuwaiti desert. Tomorrow would be ever more special. Terrible Teddy was going to be baptized. He was very excited about the upcoming experience.

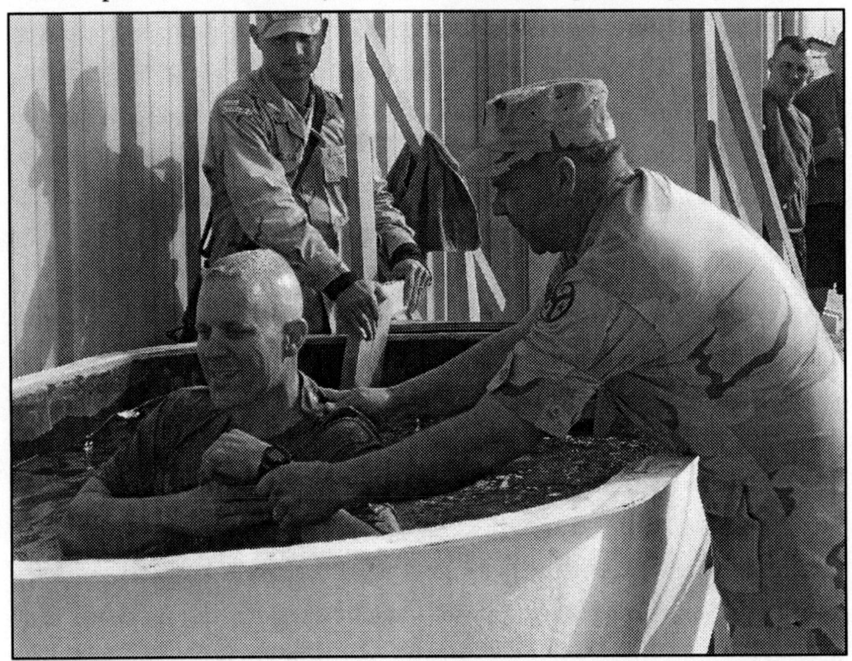

Terrible Teddy gets baptized by Regiment Chaplain

Today we worked on the Bradleys. We built shelving systems in them to transport our equipment better during the convoy into and through Iraq. Convoys were a very sensitive subject because on any given day the Armed Forces could expect to come under direct or indirect fire from small arms or RPGs or come face to face with some kind of improvised explosive device. Most recently our convoys were being assaulted by vehicle borne IEDs. Extremists were willingly driving loaded vehicles at or into our convoys. The drivers would attempt to render our vehicles inoperable and kill our soldiers. Fortunately, US convoy security was getting really good at what it was doing and in most cases the son of a bitch who was trying to detonate a bomb in our convoys would be riddled with bullets from our gunners.

The lack of a soldier's self-care for fear of being called a pussy came to a head today when Scooter asked me several times if he should go to sick call for an illness that had been kicking his ass for over a week. I tried to reassure him that the guys were adjusting to the new environment and in the process were getting some kind of bug. I would advise them to go to sick call and they wouldn't. Death Pants warned me that some of these guys would try to milk the system for everything and I was beginning to see that he was right. It almost got to the point that I felt like I needed to ask Dark Wing and the LT to make it a standing order that if anyone in our platoon was told to go to sick call, he had to go.

Enter the Dragon

December 3

A plethora of emotions ran through me today. I've had some God given encounters as well as have been annoyed beyond the realms of belief by the stupidity of some of the people who crossed the pond with me. Some of our soldiers were so lost they, like I at one time, had no clue that they were so far away from the Lord, his love, and his redemption. He was just a blurred vision of someone else's hope at the very least to some of the guys. To more than a few, He didn't exist. Most had heard the name Jesus but very few had any clue as to the peace of a relationship with Him. Every other word from their mouths was "God damn" or "fuck." It was sad. Somehow the guys had missed the memo on God's love. I was scared that in that kind of environment I might lose that which I held so dearly.

Tonight I did a tattoo of Jesus. The Lord's face was easy for me to do. He was looking up, his face with a look of expectation, waiting to return to his Father. I wanted to make Him look as good as I possibly could. The crown of thorns was another situation. I felt less of a desire to put the crown on Him. It was almost like I tried to avoid putting it on. I didn't know what it was – weird though. The tattoo turned out fine. As a matter of fact, remarks were made that I had outdone anything anyone had seen me do to date.

For the last three nights I've been wiped out having been at the zero range, guard duty, and MOUT training – Military Operations Urban Terrain. That was actually pretty cool. The training was in an area that was set up similar to a city block. Trailers and concrete buildings served as city structures. While we didn't do any live fires on that range, we were able to become familiar with how our teams operated together. We got used to each other's habits. We could correct bad habits while we were here. It was much better to get things corrected during training than have to find out during actual combat that what we've been doing was wrong. I was the fourth man in our squad's Alpha team stack. The team consisted of Pretty as team leader, Scooter as our 240 gunner, Boo Boo on the radio, and I. I outranked the others but wasn't in the leadership position because I was the medic and Pretty had been a Scout Sniper in the Corps. We worked well together while we were at the MOUT site. I was excited about what we'd be doing but I hoped that I would still be able to serve as a medic. We were getting closer to heading into Iraq. That was where we'd get tested.

Dad seemed to be really concerned about the validity of our presence in Iraq. I assumed that it had more to do with my being in combat than any real concerns about what else was happening in theater. I didn't pretend to know what my parents or any parent with sons and daughters involved in a combat situation must have been going through. Most people never find out anything about military operations but for some reason everyone seemed to know what the 278th was doing, even more so than the soldiers.

December 5

Last night was the pinnacle of stupidity within the ranks of The Deuce. A few of our more illustrious soldiers had found a way to get high. These were the guys that called themselves Section 8. They had gone to the PX and purchased compressed air cans that were normally used to dust off computers. They would inhale the compressed air and get high for a few minutes.

These guys thought they were invincible, but last night Skeletor found out that he wasn't. He went too far. After sucking down some air he passed out. He began to seize and was fighting a real battle to breathe. Crawfish, who was huffing with him, just sat there and laughed.

After a minute or so of gasping for air I told Crawfish that he needed to take care of his buddy since they were in this thing together. I wasn't going to help. He could have died and I wouldn't have done a thing. Crawfish laid Skeletor on his back and after another minute, Skeletor began to breathe somewhat normally again. Then he started to puke excessively. What was amazing was that Crawfish continued laughing at him like it was something that was no big deal.

I was fuming. I told Crawfish to roll Skeletor onto his side so that he wouldn't drown in his own vomit. He puked up enough for three people. I couldn't believe how much came out of his stomach. Finally, Crawfish started to come back down and he and Crazy took Skeletor outside to finish his vomiting. Crazy then got some sand to put on the floor to draw the vomit out of the boards.

It was a really sad situation that made me wonder about how much I could trust these guys to do the right thing, not just for themselves, but for their teammates as well. All they talked about was how many drugs they'd done, were doing, and how much they were going to do when they got home when this tour was through.

It was Sunday and I planned on going to church at nine. I needed to be in the collective presence of believers and the Lord. I needed that peace, that break in the annoying everyday routine of being in a tent with some of guys with whom I was about to go into combat. I needed the Lord's help in praying for them, that they might know peace. I needed Him to show me again how to be a light in the darkness.

Today, as I wandered around the base, I found myself pondering the little funny things that we'd seen so far since being in Kuwait. Some of the strangest were the Port-a-Shitters. They were quite useful things, but to be honest, they smelled bad, were often dirtier than the soil outside, and had all been vandalized by the bathroom prophets and poets. As one was taking care of business he could always find something to read, be it a unit's emblem or motto, a tidbit of wisdom on females and sex in general, someone's political views, or simply someone correcting someone else's spelling and grammar. It was quite entertaining.

December 6

It was now one day before we would enter Iraq. We spent the morning loading the Bradleys and drawing ammo. Each of us would go into Iraq with over 200 rounds for force protection. That wasn't a lot of ammo but it was enough for us to feel confident, especially since most of us were going in green. Already we were hearing reports of an IED hitting one of our convoys. Fortunately there were no fatalities.

We also heard that while on convoy our XO's vehicle was approached by suspected insurgents with a possible vehicle-borne IED (VBIED). He felt it was an imminent threat. His gunner directed the crew serve weapon at the approaching vehicle. When the vehicle didn't heed his warnings, he eliminated the threat.

It was lunch time and guys were loading their ammo and getting lunch from the DFAC and the satellite Subway vendor. Where else but in today's military could a soldier live in a tent but have access to Subway and Baskin Robbins? (BR had only six flavors at this vendor.) If we had have been Air Force guys, we would have lived in nice facilities as well as having all these amenities.

We had some boxing matches tonight. Most of the guys here put on the tough guy façade. I suspected we had to when we got in a situation like this, but the amount of BS coming out of these guys' mouths annoyed me. Crazy was the only one of them who had already been in combat. He had the right to talk.

Every other sentence out of their mouths was about how much ammo they were going to use on the Iraqis when we got in country and engaged the enemy. I struggled with this because I knew professional soldiers who didn't talk about combat like that. Most guys I'd met who'd actually been in combat and had actually killed another human didn't talk flamboyantly about it at all. As a matter of fact, one almost had to pry information out of them. To be quite honest, I was concerned about who would freeze and who would react as we'd been trained. It was easy to blow smoke when nobody fired back at you.

DECEMBER 9

On December 7 we flew out of Kuwait and landed in Iraq. We woke up at four that morning because some of our guys were heading out at 0530. After loading the Bradleys with last minute equipment, the crews took off and the remaining guys from the Deuce went to breakfast. Our chalk (flight) was slotted for departure at 1100. We mustered and made sure everyone was accounted for and then waited around for an hour or so until the buses showed.

As fate and good fortune would have it, while we were congregating, I ran into a former Young Life kid. He was now a Sergeant within the ranks of Howitzer Battery out of Sparta. It was good seeing him again. It was funny how small some parts of the world actually were.

We boarded the buses and I volunteered for security watch on our bus. My job was to load a magazine and ensure that no one drove at us or did something that threatened the lives of the soldiers on board. Kuwait was supposed to be a calm place so I wasn't quite sure why they thought a shooter on the bus was necessary. With no incident besides slow driving we arrived at the air base to find that we had missed our flight. Thus began the waiting game.

We waited in a holding area for people who were doing military hops. We stayed in the hooch until about 2300. We then boarded a C-130 that took us to Iraq. It was the first time I had flown in a C-130. It was cramped and loud but it could have been a lot worse. Once on the ground we grabbed our gear and then headed to what was called tent city. It was where we were to stay for the night. The tents were small but they had mattresses instead of cots. That was awesome. I'd found that the thing I missed the most was a real bed.

We woke the next day to the sounds of low flying jets. After finding out what time our chalk was to leave Anaconda, Nasty and I went to the PX and Burger King for supplies and lunch. We made it back to tent city in time to grab our gear and catch the bus over to the flight line where we were to catch a Chinook for the hop to Camp Caldwell. While I was sitting on the tarmac, I kept thinking about my cousin and how awesome it would have been if we'd have been in Iraq together. He had flown Black Hawks for the Army and Black Hawks were all over the runway.

Our Chinook ride was exciting. The pilots flew low and fast, map of the earth. One second we were speeding along and the next we were twisting and turning as our pilots practiced rolls and dives. At one point our front right gunner engaged something below us. Everyone was relaxed and enjoying the ride when out of nowhere he jumped up, manned his gun, and unloaded on some unknown target below us. The brass went flying. Great excitement because now we were in combat! Hell, yeah! We didn't know if he was test firing or shooting at a live target, but we all got off that bird knowing that we were officially in the shit.

Flying into Caldwell

****

From: *<Malcolm>*
To: *<Mom>*
Sent: *Wednesday, December 08, 2004*

Greetings from Iraq. We're here now. Just got in last night and flew into our area of operation today. It was pretty cool. We came in on helicopters. As we sat on the tarmac I kept thinking about Mike and wishing we could have done this one together. Oh well, maybe another time.

Things are good. We're out of tents and in eight man rooms in dormitory-like buildings. We even have normal beds!!!! That's the best. No cot right now. I'll probably have another address change for you in the next couple of days. Sorry.

We're waiting on the rest of our guys to show up and then we'll start our missions. Sarah Brady told me you were putting up a Mardi Gras tree for Christmas. That's cool. Wish I could see it.

December 11

We've been sitting around all day waiting on the convoy to arrive. We wanted to help square things away, carry gear, and get them parked. It's been very slow. I started reading the book *Baghdad Butcher* by Mike Jackson. It was a fictitious story about a man hired to kill Saddam Hussein. I suspected that when our guys arrived we wouldn't have quite as much time to sit around and read.

Scooter liked watching movies. I've noticed that anytime a television screen was on and he was around, he was glued to it and it would take him a few minutes to acknowledge whoever was trying to get his attention. It was amazing to me how much our younger generation needed the TV to give them something on which to focus. They didn't read much, just watched TV or played video games on the TV.

Boo Boo continued to work on his computer. He was trying to get all his downloaded music organized so he could be our DJ. The rest of us lay around and read. All we heard were the sounds of Chinooks in the distance dropping off people.

Scooter told us that he wanted to have a career as a millionaire serial killer. He was a young kid, only 20, and had a tendency to get on our nerves but he was all right. He didn't need to get caught up in all the BS of trying to be a tough guy. He didn't need to try to impress anyone. That was one of the toughest things to deal with for me. I was all about being cocky, but when a person became arrogant in his words and actions that, to me, was a sure sign that I didn't need to be around that person. I didn't need to become like that person and from what I'd been told my whole life, a person became who he hung out with. I didn't want that for Scooter either.

DECEMBER 12

Today I met the unit Doc to whom we'd be turning over duties. He seemed like a good guy. He knew I was a medic filling an infantry slot; but in order for him to be cleared to leave Caldwell, he had to turn his duties over to another qualified medic in a process called RIP.

With our Bradley guys here, we were the Double Deuce again. We traveled in a pack. It was funny to see everyone huddled around Dark Wing. I didn't cluster with them when we walked to various locations because they appeared to be like a pack of puppies trying to fight for mama's last nipple. Crazy was the biggest pup of the bunch.

We were slotted to start doing missions within the next few days. I was going to be shadowing the medics from the old unit on Wednesday. We had been told that there was some activity in our area of operation but we weren't sure what to expect from the insurgency. They didn't operate on a set time pattern. We knew that they had a tendency to hit the same spots simply because those spots had worked in the past. From information put out in an intelligence report, I learned that most of the attacks in our AO were being done during the phase of the full moon. The insurgents needed more light to operate effectively at night.

Three of our guys who brought the Bradley into Iraq

A lot of the guys have been trying to call home and others were doing things on their computers. Scooter was plugging away at some computer game. He finally got his computer working thanks to Tater allowing him to use his power cord. It was funny to watch him and how engaged he was in the game. That was his escape from the environment that we were in. A lot of kids back home were like that. The dysfunction of American families has made computer gaming software a viable escape for millions of youngsters. It gave soldiers something to do as well.

DECEMBER 13

Today we met our new platoon leader, Big Red. He was a former enlisted man and was a bit long winded. In civilian life he was a trauma nurse. I hoped that I would get the chance to pick his brain a bit and see what he felt would be a good training program for the individual soldier to know what to do in the case of emergency while on patrol or watch. I could teach them what I knew from Whiskey training; but he worked in blood every day so I'm sure he had some ideas that would be high speed. He was also a Christian. That was a relief for me to have another person, especially one in a leadership position, who was a practicing Christian. I was encouraged.

Dark Wing was out on patrol riding right seat with his counterpart. After a few hours Dark Wing was back. He's been telling us about the town of Balad Ruz. He said that the 1st ID (actually the 130th from Hickory, North Carolina), whom we were relieving, was very professional and did good work. Our guys kept asking if there was fighting today but Dark Wing said nothing had gone on today except their patrol checked out some places where IEDs and engagements had occurred in the past. There was a culvert system where the Bradleys test fired their weapons systems. The crews did it there because explosives had been found there some time ago. It was a good procedure to ensure the safety of the passage of US convoys.

We learned something about the Iraqi culture today. It seemed that relationships between men were commonplace. Of course that seemed really weird to us. The men gave a hand signal to one another by rubbing their index fingers together to let the other men know that they were interested in consummating their relationships. Apparently they only have sexual relationships with women for the purpose of procreation.

Big Red called out Pretty today because he said that if an American was attractive in any way, the Iraqi men would probably be hitting on him quickly. That got Pretty nervous and brought a round of laughter from everyone else in formation. We came back to the room later and started a pool to bet on who would be the first approached by an Iraqi and Pretty was the hands down favorite for man love.

December 14

Boo Boo has been telling us the intricacies of the male stripper industry. Everyone has been ribbing him about it. It was quite funny to know that one of our own used to take his clothes off to entertain other men so that he could earn a quick buck. Usually when I heard about a girl who had done that, I'd get sad because it was almost like she'd lost her innocence; but something about Boo Boo telling his story was absolutely hysterical. I suspected it was because he tried to justify the ends by the means.

From 0900 to 1500 today we added reactive armor to our Bradleys. I didn't know what the hell it was until we started busting hump to put it on the Bradleys. It was a shape charge of C4 that reacted when the velocity and impact of an explosion set the C4 off. The reactive armor insulates a vehicle's armor so that the projectile and its explosion don't penetrate that vehicle. In fact, when the projectile impacts, ideally, the reactive armor cells explode outward, away from the vehicle, forcing the initiating explosion out with it.

The way our barracks were set up, we could see directly into one of the other barracks. Some of the guys were checking out the females who were across the courtyard. Bong Water was absolutely hysterical. He would always get caught because he'd never turn out the room light. Another funny thing was that there were some girls on another floor who were peeping over at our barracks. So the game began. Who would be bold enough to make the next move when we figured out who was who?

DECEMBER 15

Tonight a few of us were scheduled to ride right seat on a raid. I wondered how it would go. The Old Hickory unit seemed to know what they were doing and hopefully our zeal to be out on a mission wouldn't get in their way. I worried sometimes that these guys talked too much about killing. They still thought they were invincible. From my perspective, they were underestimating those with whom we could be in conflict.

Too many reports had come in about the calmness in the area and the guys were looking at the situation as if they were bummed out that nothing was happening. Granted, I wanted to do my job, but I wasn't looking forward to day in and day out waging battle. Nothing about that thought intrigued me. If we came under attack, I felt prepared to fight and would with every bit of life that was in me. If we did a raid, I felt like we had the upper hand and we could kick anybody's ass that got in our way. But in no way did I want to find myself in the midst of a sustained skirmish. That was for Hollywood. I wasn't Hollywood.

Tre Deuce prior to roll out on patrol

Dark Wing began talking about his mischievous adventures in high school again. It sounded like he and Pretty could have been in chemistry class with me. They were talking about blowing up things with oxygen, Bunsen burners, mercury, and everything else to drive the teachers crazy and get their friends to laugh. Ms. Gunn was one of Dark Wing's teachers. She was old school. She could slap the student around and not have to worry about some parent crying to a lawyer. Good old discipline.

There was also the story of the economics teacher who would lose his mind in class when the girls would flash him their boobs or were not wearing panties, only short skirts, and they'd flash him their privates just to see his stammered reaction.

Finally, there was the story of Moose. He was the football coach who could paddle so hard that Dark Wing had to walk across the gym floor and up the bleachers before he could catch his breath. I had a fifth grade teacher who lit me up like that one day. He caught me acting up in the hall after P.E., gave me the "Your butt's mine" signal, and then fired me up with a few licks from his board of education. He was a great teacher and I admire him still.

****

From: *<Malcolm>*
To: *<Mom>*
Sent: *Wednesday, December 15, 2004*

Yes, I got a letter from you dated on the 29th of November. I guess I got it on the 14th or something like that. It's the 16th now. Three more days and I'm 35. We went on our first patrol tonight. It went smoothly but I'm glad we got it out of the way. Anticipation gets crazy at times. The guys have been chompin' at the bit to get out and do what it is we do. All right, here's the address for while I'm here at Caldwell. This is it (I hope).

Malcolm Rios
A Troop 1/278 RCT
OIF III Camp Caldwell
APO AE 09374

December 16

Last night I went on my first patrol. We rolled out at 2000 for Balad Ruz. I rode right seat with an Old Hickory medic. He was good about letting me in on what had been going down in the Balad Ruz area. We rode in on the CO's Bradley and dismounted when we got to the Iraqi Police station. It was surreal, that first time, unloading from the Bradley with weapons at the ready.

Prior to leaving the track line and while on the way out to Balad Ruz, I prayed to God. I wondered how many other soldiers did that before a patrol. I prayed a lot while we were patrolling too, praying for our alertness and safety. Once in the town, our mission was to "bust out" three Iraqi National Guardsmen who had been jailed for the death of a lady in Balad Ruz. Apparently they had been falsely accused. I assumed they were released without much incident because while the deal was going down, Doc and I were outside doing security on the Brads and never heard a thing.

After the ING were turned over, we loaded back up and drove a few blocks down the street to the main traffic circle. We turned right, entered the hospital area, dismounted again, and began patrolling the side streets and alleyways for the next few hours. Part of our mission was to do soft knocks. A soft knock was when we knocked on a door and allowed the Iraqis who lived in the house time to answer the door before we entered. A raid or hard knock was when we kicked in the doors and entered because we had hard evidence that there was a threat to our security. There was no forewarning in hard knocks. What usually happened was that we would kick in a door, search for weapons caches or illegal contraband, whatever was illegal to have at that time, and if things were found, we would take them. If nothing was found, we would carefully back our way out because one never knew whom one might have pissed off.

The soft knocks went off without a hitch, and for the most part, all we saw in the streets were dogs. There were a lot roaming the dirty streets. Apparently the Iraqi people didn't keep them as pets. They just wandered around, barked at soldiers, and howled as the artillery illumination flares were shot over the town.

The patrol ended at 1130. We drove back to Caldwell, cleared our weapons, unloaded the Bradleys, and then hit midrats (midnight rations). I took the opportunity to run down to the Internet center to check and send emails. I found out that my cousin, Mike, had been promoted to Lieutenant Colonel. He was working at the Pentagon and has always been a great source of inspiration for me.

December 17

Today the guys asked me if I thought of them poorly. I suspected that they had finally picked up on the fact that I wasn't impressed with their shit talking. They asked me if I hated them. Did I? It would take me some time to sort through my thoughts and feelings on that one, but yes, I was leaning that way. While 80% of the conversations that we'd had thus far were about drugs and treating women like dogs, I couldn't say that I truly hated them; but they were really starting to get on my nerves.

I had been part of youth outreach for the last six years, something that was so positive. The conversations we were having now were on the opposite spectrum of what my daily conversations were in those ministry teams. I missed intelligent, stimulating conversation. The guys asked me if I was scared of them. No, they weren't so bad that they scared me. The only being I was scared of was God. We all knew I was different from the rest of our group. I wasn't any better than they, just different. There were a few other guys in our platoon who were as old or older but they seemed to be more passive in their opinions on things. Passivism has never been a favorite characteristic of mine.

December 18

Crowfoot just blew up at Boo Boo. Dark Wing came in and said something about Boo Boo being the last one out of bed. He was dragging ass when he should have been up getting his duffle bags. They had been missing for ten days. We'd been told a few times that they were on post, but every time we'd gone to find them we went on a wild goose chase. None of the people involved with getting our bags to us had bothered to get good information prior to telling us that the bags were on post. On our end, we were stupid in that we didn't verify the information prior to moving out to find them.

I was listening to the Metallica song "Of Wolf and Man." Their music has always been a good way for me to escape from life. In Iraq, they helped me block out some of what was going on. Some of the guys were discussing how poorly their recruiters had set up their records.

Others were talking about Christianity vs. Satanism and how the Satanic bible made no mention of sacrifice. The Christian Bible was full of sacrifices. As a matter of fact the Christian Bible had stories for just about every topic from sacrifice to adultery to incest to murder. The common theme throughout the Book was redemption through love – a loving relationship between God and His people and their everyday struggle with that faith and love.

DECEMBER 19

Last night we had fight club. Tater and Scooter boxed for 28 seconds. They came at each other like Rocky and Apollo. Scooter groveled away with a bloody nose and Tater finished the bout with a dislocated shoulder. His arm socket was just a little bump on top of his shoulder. His arm was dangling like a limp noodle. When the match was called, Tater came over, wincing in pain.

This was my first dislocated shoulder and my first injury in the zone. Granted, it wouldn't get Tater a Purple Heart, but it sure gave us something to talk about for the next few weeks. He was feeling nauseous. I needed him to remain standing so I could relocate the arm. I tried tying some cloth around his wrist to help pull but it didn't work so I just pulled his arm manually. It made a bizarre squish sound when I pulled it. Tater went from a staggered shock stage to a cold, clammy, profusely sweating stage. I set his shoulder and he took some time to lie down.

After taking care of Tater, Crazy and I were up for our match. We laced up our gloves and then went at it. Each period was one minute in length. Despite our poor conditioning, we had a pretty good brawl. Crazy came in low. He was trying to bulldog me. I allowed him to come in by stepping back and waiting for the opportunity to get some quality hits in rather than just flailing like so many others. He hit me a few good times and I did the same to him. After fighting another person one feels like he knows that person a lot better regardless of who wins. I felt comfortable with Crazy now and knew he'd be one of the ones I could count on if or when we ever got into the mix.

Tre Deuce in all its glory

We were just called up to go into town. The A Team got locked and loaded and was out at the Bradley when we were given the call to stand down. Apparently a convoy that had gone through Balad Ruz earlier had fired some warning shots. The Iraqi National Guard was with the convoy and fired warning shots too. The shots soon turned into an engagement that left four Iraqis wounded and one dead. When we were called, the word on the radio was that shots were still being fired. Later in the evening it was revealed that the last shots had been fired in celebration.

The guys in the Double Deuce have been talking death since Camp Shelby. It wasn't a passing thing. Most of the younger guys, the guys who called themselves Section 8, kept saying that they were looking forward to killing someone. In my eyes, they were talking that trash to pump themselves and their buddies up - macho bullshit. Don't get me wrong, I'd thought about killing people my whole life. I thought that was what happened when one become a warrior. I had to think about killing because at some point, I might actually be forced to do it. And yeah, I've thought about killing even outside the military ranks. To quote a Jane's Addiction lyric, "Some people should die. That's just unconscious knowledge."

I had yet to look forward to killing somebody in Iraq. I wasn't angry enough. I had no problem doing it if and when the time and place called for it, but it wasn't my thing to talk about how I was going to do it every second of the day. Some were overly zealous about taking another man's life. The thrill of the kill wasn't as glorious as it was cracked up to be but they had no idea of that yet. Killing was something that had to happen in war. It was inevitable, but bragging about it was annoying, especially when a person hadn't proven himself in battle. I wondered if some of these guys' lives had gotten so bad that their only hope for retribution was in taking out their frustration and aggression by taking another man's life.

I reenlisted to defend this nation from the jackasses who would try to destroy our way of life. If that meant killing them then I was more than okay with taking their lives. If we didn't stop them they would certainly strike somewhere else in this world, disrupting the lives of innocent people in the name of terror and Allah. Some people were not willing to defend freedom in that manner; violence, in their eyes, could not beget violence and still be a good thing. Some willing people, usually old vets, were not able to put their lives on the line to defend our nation and the lives of all people seeking to live in peace. I was. I knew that there was a better than average chance that this journal would never have the words, "We all made it home alive, including me." But that did not mean I had to or would be excited about the "opportunity" to take another man's life.

DECEMBER 20

We sat around for most of the day either in the room or at the chow hall. No one here had a clue it was my birthday but I did receive some nice emails from back in the States. It was amazing to me how the Internet helped us stay so updated this far away from home. I couldn't fathom how guys in all the other wars did it. I guess the snail mail gave them a huge sense of anticipation. I didn't worry about getting snail mail too much because we had such liberal access to the computers and Internet downstairs.

I realized that I wasn't a die hard Double Deuce man even though I had the brand. I state that because I didn't feel any closer to these guys than the day I walked down to meet them at Shelby. I still felt like I was always on the outside of things. I was not a drug user or a big whoring drunk (any more).

Sure, I'd done my time in the bars and chased skirt with great success. I'd been there and done that and wasn't claiming to be any better than they. It was just that I felt so much heartbreak when I listened to their conversations and that's what the majority of their conversations were all about. They had very little depth. I didn't know how to put cracks in their walls much less break them down. Sometimes I felt like there was hope but when they were all together they would feed off each other. I was certain that not having been at Camp Shelby with them for the duration of training I was behind the curve on things. That had a lot to do with it, but most of it had to do with where we were coming from in life. Our lives were just different.

I'd also come to realize that I'd never get promoted while in Iraq. That was something that was hard for me to fathom. The possibility of promotion in any endeavor made me strive harder and work better. I had to get it into my head that I still had to work harder than these guys even though I wouldn't get promoted. Dark Wing was already talking to some of the guys about promotion. Part of me believed it was because they were his boys and he was looking out for them. The other part of me said, "Hell, if they deserve it, give it to them." The question in the back of my mind, though, was did they really deserve it?

That made me question whether or not I deserved it too. I was a medic but I'd been filling an infantryman's slot while doing medic work for our platoon. Every patrol we went on, despite having another medic with us, I carried my aid bag and would have taken over the situation had one arisen. I was the senior medic when we were in I Troop. I had the duty of ensuring our medics and Combat Life Savers were equipped with the supplies they needed for first aid on the battlefield and off. I believed I did my job well. I was frustrated that I would not have the opportunity to go up for promotion.

I've started my workouts. I wanted to start so that when I got back I would be better prepared to submit a Special Forces package and take any physical exams. I did not want to be in this slot any longer after our time in Iraq if there wasn't a chance of promotion. I wanted to go either SF or reapply for OCS.

Special Operations was something I'd always dreamed of, working within the ranks of the elite units. When I joined the Navy, that's what I wanted to do with the SEALs. I busted my ass to take and pass the entrance exams. The detailer I was under axed my request to go to BUD/S. He said that my unit, the *USS Tortuga (LSD-46)* was undermanned and needed me there more than the Teams needed me. I took the test a few more times and kept trying to get in but the same response always came back to me.

I submitted a package late in my enlistment to apply for Special Boat Unit duties in Panama. It would be the deciding factor as to whether or not I would stay in the Navy. In my detailer's infinite wisdom, he made the same call as when I submitted my SEAL packages. Although I was fully qualified and my commanding officer had given me the "Okay," I wasn't allowed to go because some other person, whom I'd never met, thought I would be a better asset where I was currently stationed.

Prior to coming to Iraq, I was enrolled in Officer Candidate School. I dropped out of OCS when our unit was activated so that I could go to war. I believed in what we were doing in the war on terrorism. After dropping from the course, I was sent to Fort Sam Houston for medic training. At enlistment I chose the occupational specialty of 91W, medic, because all units needed medics. Obtaining the MOS qualification would allow me to have a lot of options if I chose to transfer to another unit.

DECEMBER 21

Last night we were slotted for a patrol of Balad Ruz. After getting our gear together and massing at the Bradley parking area, we were told that our mission had been changed. Our A Team and Bradley crew, along with four other vehicles and the soldiers operating them, 24 Joes, would be going into Balad Ruz to recover … a generator. It was funny how things changed sometimes and what a priority became. We got behind the curve a bit because our unit had overbooked the Hummers we were supposed to be using. For some odd reason, we had to "hot seat" with another element of our unit that was still on patrol while we were massing for ours. It set us back about an hour and by the time the other element showed up, our guys were restless and chomping at the bit to get out of the wire.

Most people never thought about the cold that was part of a desert setting prior to going into the desert. Although we had been told that this area of Iraq would hit 150 degrees in the summer, at night the temperatures had been in the low 30's. We dressed with poly-pro undergarments, Gortex outer shells, and gloves. I assumed that a lot of the guys who were sick were that way because they would ride in the backs of the vehicles in this kind of weather at speeds of 35 to 45 miles an hour; the wind-chill making it that much worse.

We arrived in town and got to see some of our guys who were sent to man the JCC. They seemed to be doing fine because they had a satellite dish to watch some scandalous movies that supposedly we weren't "allowed" to watch in this country. Sometimes we just had to make the best of the situation in which we found ourselves. Our guys made good use of the porn channels that were being piped into our position.

In the town of Balad Ruz, each family was allowed one firearm that they must keep in their house. They were no longer allowed to carry the weapons in town. The only personnel authorized to carry arms were the Iraqi Police and the Iraqi National Guard. Most of those guys carried AK-47's.

We came back to Caldwell at 2200, in time for midrats. A few of us went down and had chow before we all hit the rack for a 0400 wakeup. We started the day with another patrol. The Bradley led the column through town. Once we arrived at the western boundary of our AO, we dismounted to fire our weapons. We also had to take off some concertina wire that had gotten caught in the left track of the Bradley. After we removed the wire, we rolled back into town and began our foot patrol.

This patrol was the first one that most of us had done during daylight hours. We were able to see a lot of the town that we had only seen in the shadows of night. It was a dirty town: calling it piss poor would be giving it too much wealth. Trash littered the streets. Sewage ran through channels dug out by the people from their houses into the streets and down to collection areas that smelled as horrible as the feces running into them. Most garbage was simply thrown out the door into the street. Packs of dogs ran wild through the town as did small herds of cattle. Every once in a while we would see a random cow rifling through the garbage in hopes that it would find something wholesome for its diet.

We saw a lot of children on our patrol. Most were heading to school and would say, "Hello, Mister" as we walked by. One little girl caught our attention because she lived in the less developed part of town (as if there was a more developed part) and came out to greet us as soon as we offloaded. It was still in the 30's and out she walked with no shoes! It was so sad to see that kind of poverty. It was sad that morons out there would be willing to die to keep that lifestyle dominant in Iraq. Oppression is evil. If anything, the Iraqis should look to their children for hope. Our world needs to reevaluate where its priorities are. There has to be a way to give these people more hope without having to force them into hope at the front of a gun.

The night's patrol left without me. I sat this one out so that one of our Bradley drivers could go out on foot with our team. Most of his time was spent in the hatch shuttling us around and I assumed he just wanted a change of scenery. Unfortunately for him, the guys weren't slotted to dismount so his perception would be from the back of the Bradley as opposed to the front.

The guys just got back in from the night's mission. All went well from what I gathered. When Pretty came in I was playing guitar and he, Crowfoot, Boo Boo, and I began singing *The Dukes of Hazard* and *The Jeffersons*. Those themes were ones that everyone knew the words to – well almost everyone. Boo Boo didn't know the words to *The Dukes of Hazard*. We razzed him for a bit but he made up for it on a few other songs.

We also read in the paper today that seven Marines died in Fallujah on Sunday. Pretty and Crowfoot, being former Marines, both felt a tinge of sadness for their brothers. We were all sad. It was a crazy war, one where we would go out to the end of town just to be shooting, while in Fallujah brothers from other units were being killed in firefights and by roadside bombs.

December 22

Tonight we got a special surprise. Andy, Korn Head, and Crawfish showed up at Caldwell. They were on their way from Bernstein to Baghdad and had been telling us about their adventures so far in their part of the war. They were telling us how messed up some of the things were at their post. They seemed to be having issues with our old LT.

It was crazy how often the conversations were about how bad everyone else was compared to the enlisted men in the Deuce. My issue with that was that if we were that good, why didn't we take it upon ourselves to help these guys out instead of bitchin' about how bad they were? We could have set an example for them and not whined about it all the time. I guess I was just getting tired of hearing all the bad remarks that the guys were making. I was tired of dealing with the attitudes and the arrogance of some of our guys.

Nasty - the Double Deuce's #1 gunner and SAW man

Our patrol was pretty interesting today. I played gunner on the Bradley. We traveled south of town to see how far we could go until we lost radio contact. The three vehicle convoy only went a few miles south of Balad Ruz. We then ran a canal and dismounted. Our patrol interacted with a few people, mostly kids. The landscapes outside the city weren't much different from the inside, sand and trash everywhere. A lot of shepherds were out with their flocks.

I did get to shoot the 25mm on our route west. As I said before, we would clear the culvert system midway between our base and Balad

Ruz. This time I was able to "pull the trigger." It was actually a button, but it had the same effect. It fired the weapon. I did not know how to line up the sights so when I shot, the projectile went over the berm. I'm sure it scared the hell out of everybody in the vicinity where it struck, but Dark Wing and I got a good laugh out of it. He never let me gun again though on the Bradley.

December 23

We spent another day in the town of Balad Ruz. Wake up was at 0400 and after we suited up, we were at the Bradleys by 0445. Our plan was to give some school supplies to a local teacher and then foot patrol for a while. We were doing this because the school's headmaster had been killed when his vehicle erratically approached a US convoy. After failed attempts to wave him off, the gunner of the trail vehicle lit him up. The incident had left a bad taste in the mouths of the Iraqi people in our area. We wanted to rectify an unfortunate situation.

We left the teacher's house and ventured down some side streets. Swarms of kids assaulted us, wanting goodies. We handed out candy, pens, and power bars. Our patrol continued for some time but eventually we loaded up to head to another location to do a soft knock and search. The search turned up nothing but more kids. As was the norm, Big Red and Atlas had bags of candy for them.

On our way back to base, we were passing through town and saw a guy in the crowd on Market Street who was covered with blood. We turned the convoy around to try and see what was going on but could not find him. He could have been slaughtering chickens or sheep or killing people. We didn't know and since he didn't care to be found, we headed back to Caldwell.

December 24

Twas the night before Christmas and all through the base
Not a soldier was back home, we were stuck in this place.
Maintenance day and what was our plan?
To work on the Bradleys and bust ass for the man.
We took off the old tracks and linked up the new.

By five in the evening our work was all through.
Initially, we had issues. The pins seemed to all stick.
Then Jonesie showed us something that turned out to be the trick.
He aligned the ridge and the groove so that the pin
With a tap from a hammer would easily slide in
With little or no effort the lesson, a perk
We picked out our partners then got back to work.
A few were the track draggers. A few used the zapper.
A few used the hammer, the drift pin, and crapper.
Big Red was there and he even bought drinks.
He bought candy for all of our Mung Ho high jinks.
No one could believe that we'd finish by dark
But at two we decided to leave them our mark.
We set up a race to see who could work faster.
One team on each track with Dark Wing as master.
He kept the clock ticking as Crazy dragged wildly
With Pretty, sweet Pretty, who pulled oh so mildly.
Then team number one worked against team number two
And sooner than later our work was all through.
Priest and Tater were both in driving mode
So they parked our old Bradleys by the side of the road.
We cleaned up and washed off the grease from our hands
Then we walked down the road to eat dinner with friends.
Now we're all back here and I've taken a shower.
Old Crazy was there and could have stayed for an hour.
So this holiday season when the lights are a-glow
Remember the Deuce. Merry Christmas, Mung Ho!

December 25

I called home today for the first time in a month. I couldn't discuss the details of what our operations were at Caldwell. I tried not to bitch too much about how some of the days have gone while trapped in this barracks with some of the guys. This afternoon I dared the guys to see if the entire room could go ten minutes without cussing. It was less than two seconds and Scooter dropped the "F Bomb."

The cooks hooked us up with a great Christmas meal. We had the works: turkey, ham, roast, dressing, rolls, shrimp, pasta, you name it. They had also set the chow hall up in holiday display. It was a good escape from the confines of the room and their efforts made being away from home a little more bearable in spite of an all day rain.

December 29

The past few days have been a swirl of boredom and excitement. On the 26th our fire team was assigned to Joint Command Center (JCC, sometimes call the Joint Operations Center) duty. Our mission was to go there and stand guard for 48 hours. It wasn't too bad but it did get boring. We stood four hours on and four hours off during daylight hours and then two hour rotations throughout the night. We had a 50 caliber machine gun on the front roof and two 240 mm positions on the side and back. The sniper rifle moved between positions depending on the hour of the watch. The radio center also had to be manned for incoming and outgoing traffic.

While standing watch outside during daylight hours we were able to see what was going on in the city and interact with people. Some of the Iraqi kids would come by at different hours of the day and say, "Hey, Mister. Chicken, kabob, two Pepsi. $5." It sounded pretty funny in their accents. They were driven by a desire to get something from the Americans. They would make money from the transaction or they would get a handful of candy thrown at them.

On my last watch that rotation, I was manning the 240 and watching a bunch of kids have a water bottle war. About ten kids came out of nowhere and found a bunch of discarded water bottles. They filled them halfway with dirt or muddy water and then commenced throwing them at one another in the hopes of nailing the other team and not getting hit themselves. Some of the older guys could throw pretty well, but there were a few of the younger, smaller kids who couldn't hit the broad side of a barn. It was interesting to watch them dodge and run back and forth trying to gain the advantage on the other boys.

On a normal rotation we would stand our watches and then sleep when the other team took over. At night we'd stand watch on the third floor of the JCC. This was where the 50 caliber machine gun was situated. It got very cold up there because of the north wind, so we would dump some kerosene on dirt, let it soak, and then light a fire to keep warm. It was fine until we ran out of kerosene and items to burn.

Another one of the interesting things about JCC duty was the availability of porn. When we were deployed we were told that we couldn't bring any form of porn, not even pictures that guys had of their wives. But most guys who wanted it still found a way to get it here. They were just very discreet in their viewing of it. At the JCC, we had internet and satellite. The ING and IP would be checking out porn the entire time they were in the building. The satellite was a different ballgame. We had the hookup in the radio room. There were about five channels which did nothing but advertise for video and Internet porn. Ladies would do strip acts for the camera in hopes that whoever was viewing would call in and pay money to chat or view more of what they had to offer.

One channel in particular had girls who would be on a phone with a remote for the camera so they could zoom in their shots. The girl who was on the camera at the time would act like she was receiving a phone call from a viewer and then begin to do a phone sex / strip act for the viewing audience. The thing that was funniest to me was that they all seemed to be on the same black couch with the same yellow phone. They'd look so seductive and then begin to touch and rub themselves while enticing the viewer to make the call. The longer the caller stayed on the phone the more clothes the temptress would shed and the more she would do self-gratification. This would go on for about 15 minutes or until she was completely naked and pretending to masturbate. Though the channel never showed actual penetration, the girl would certainly make one believe she was about to go to unexplored levels of self-pleasure.

I watched this for a number of reasons: there were only two or three English speaking channels on the tube, one of which was BBC and was busy covering the earthquake and tsunami which devastated southeast Asia; some of the plots in the striptease were very corny like the one with the girl and the Ferrari so they were funny to watch; and finally, after being in this country for almost a month and not having seen any remotely attractive women, it was more of a desire to fantasize about that which I did not have. The girls just offered us an escape from the whole deal of being with a bunch of guys. Granted, that probably wasn't the most decent way to do it, but we learned to make do with what was available.

We left the JCC and returned to Caldwell to move from our barracks into Conex Housing Units (CHU). The CHUs were what we had been looking forward to the whole time we had been in country. They were two-man units so they offered a much needed break from multi-person housing. Boo Boo's side of the CHU was very organized with his single bed, computer nightstand, desk, and black box while mine remained in a constant state of disarray.

On the night of the 28th we went out for another patrol at 1900. Our intent was to drive to the eastern most section of Balad Ruz, dismount, and then begin foot patrols. As our convoy of four hummers pulled off Route C1 onto a dirt road, we went blackout. We pulled in and then turned our lights off so that the surrounding threats wouldn't see the number of men dismounting. Our last vehicle idled on a culvert and was waiting to pull in to the dismount site when the culvert gave way and the Hummer began to roll. Fortunately the back passengers were very alert and they pulled the gunner down through the hatch.

Immediately the word came over the radio that a medic was needed. My immediate response was, "For what?" Then word of the rollover exploded over the airwaves. I jumped out of our vehicle and ran to where I saw the Hummer on its roof and smoke coming up from the engine. My first thought was, "Oh God, don't let it blow up." I guess I've watched too many movies but I was thinking that where there was smoke, there was fire. If there was fire then there was a possibility that something could explode, either fuel or ordinance. If there was no fire then we'd have a better chance of getting everyone out alive if there were any survivors and we wouldn't be as rushed to extricate them.

Rollover! Rollover! Rollover!

I slid down the ditch after tossing my weapon and began to assess the situation inside. Two other guys had made it to the other side of the vehicle and began getting the driver and the driver's side passenger out of the vehicle. Three guys were still inside. The Hummer was still running so we shut it down. Shrek, who had been riding shotgun, was trapped beneath his seat and said that his leg was jammed. Adrenaline and fear were consuming him and my first priority was to calm him down. He was scared for his men and himself. The gunner, who had been pulled in by Atlas and Worm, was on his back trapped underneath the two other soldiers, his head and upper torso trapped beneath the radio unit in the vehicle. Atlas, who had been riding in the back right passenger seat, was in the oddest position of all. He was on his back with his legs crunched above his head similar to when a weightlifter does incline squats. The jarring had jammed him and he was complaining of back pain.

Shrek looked the easiest to get out so I moved to him first. I attempted pulling him by his IBA vest. When this proved futile due to the seat wedging him in, we began to work at getting the chair dislodged. Once this was done, we were able to get the seat in a reclining position which allowed us to get Shrek out of his IBA vest. This proved to be the trick, and after a few minutes of further negotiation he emerged from the hull of the wreck. He had no visible injuries but was ordered to the rear so we could finish working on the rest of his crew.

Next we dealt with Atlas. In his situation, the first thing we had to do was get his legs down so that he would be supine. I asked him if he felt any major pains in his back or legs. He told me he did have some pain in his lower back so OB, the other medic, held his head and neck while I slowly helped Atlas lower each leg. In doing this I had to extend his legs over Vinnie, the gunner, who was still underneath the radio unit. After getting his legs down we began the process of moving him out of the vehicle.

With OB at his head and me on his legs we did the old "1, 2, 3" and slowly moved him out the left side of the vehicle. A couple of other guys came down to assist OB with the extrication and support of Atlas. The soldiers carried him up the bank to a more secure area where OB could continue medical treatment. While that was going on I continued to work with Vinnie. We worked to get him untangled from the radio wires and everything else that was keeping him constrained. He was able to sit up after being pulled out from under the radio unit. He then climbed out of the vehicle with a little assistance. He walked away from the incident with only a separated shoulder.

After ensuring everyone was safely out of the vehicle, a few things began to happen. We began to remove all sensitive items from the Hummer which included any gear the soldiers had had on them at the time of the accident. We also removed all ammo from the vehicle. The 50 cal. was buried deep in the embankment so all we could do was ensure that the chamber was clear of any ammunition. We couldn't get the radio out because of the lock that secured it into the dash. Diesel continued to pour into the Hummer from the tank. OB was assessing Atlas. Big Red called in a 9 Line medevac and coordinated a sensitivity item check. The rest of the guys pulled security.

After doing my part to get everyone out of the Hummer in one piece, I went to assist OB with Atlas. We placed him on a stretcher, secured him, and then tried to make him as comfortable as possible. OB was carrying a space blanket so we were able to put it over Atlas to keep him warm. He was complaining about tingling in his legs and that his hands were extremely cold. We unlaced his left boot and I gave him my gloves and my DCU top to help him warm up. Fortunately the adrenaline in me was enough to keep me warm until the ambulance came.

The ambulance, FLA, came under fire when it turned onto the road we were near. We don't know why but no one in their convoy returned fire. Upon its arrival we loaded Atlas. Crowfoot and Pretty helped us out. They became my "go to" guys whenever a situation arose that I needed stretcher bearers.

The FLA headed back to Caldwell with Atlas, Vinnie, and their security escort. Ten minutes later a Black Hawk touched down. The crew was flying blackout, no lights and with night vision goggles. It was really cool to hear the chopper in the distance and then see it swoop in like a menacing black phantom, ready to devour what was on the ground. The pilot sat her down in a field about 100 meters from our position, idled for about five minutes, then took off again once he got the word that the patients were already being shuttled to Caldwell on the ground.

We maintained a degree of security as we waited for a tow vehicle to come to our position and recover our Hummer. As we were doing this our south security stopped an approaching vehicle. The man would not do as he was instructed. Pretty engaged him with force. The man finally understood that he was not coming through our position and turned around, fleeing the scene much faster than when he had approached.

Tonight we went on another patrol in town. I walked point past long walls, separating the city housing from the palm groves, and into dead ends that made us laugh at the thought of trapping the insurgency in a situation where we could slaughter them. Never mind the fact that they could have done the same thing to us.

While on the patrol we encountered a young man carrying food back to his residence from a friend's house. As point man I thought it should be my duty to ensure that the food was safe for anyone else who wanted to try some. He gave me a little cookie thing which turned out to be something more similar to cornbread.

Growing Pains

January 1, 2005

On New Year's Eve I thought we were in for a rather routine, boring day here on Caldwell. My CHU-mate and I were not scheduled for any patrols. We had planned to sit in and watch a couple of movies. What we really needed to do was get into the gym. I would do it. I had to. Living on Caldwell would either make me fat or I'd get so ripped because in my down time, there wasn't anything else to do as productive.

Around 1900 on New Year's Eve Old Crowfoot came by to tell us that we were slotted to go on the morning mission that the company was up in arms about. When Dark Wing told us about the mission that morning he said the 1900 meeting was a company rehearsal. Immediately we asked him what targets we were hitting because we didn't go out company strength unless it was a raid or search and seizure. He told us the potential hard target was a vehicle-borne improvised explosive device-making building.

The mission was simple: three coalition elements would ride into town at sunrise, secure the suspected area, enter the subject's house, and detain him and whoever else needed to be, recover any evidence, and get back out with everybody in one piece. For this particular mission I was driving one of the five Hummers that made up A Battery's part of the convoy. An Operational Detachment Alpha fire team, the Special Forces guys, spearheaded the mission with about 15 Iraqi National Guard rounding out our group.

We entered Balad Ruz and found the street where the first two target houses were. Four of the Hummers and their compliments set up perimeter security on the corners of the town block. The ODA team then did a breach and entry with the ING acting as back up. After clearing the houses one of the ODA guys came by and told me to follow him in my Hummer. From there we went to the final target house involved in our search.

Once there I parked to block the road from anyone coming into or leaving our specific area from the North. My guys dismounted and set up a perimeter to contain anyone coming out of the house. The ODA guys then breached the back door. The ING joined our guys and helped secure the perimeter. It was the first time we had worked with the Iraqis on an operation. They risked their lives just by being involved in the ING. We would often find them wearing hoods to hide their faces so that none of their neighbors would know who they were, thus eliminating the threat of someone killing them for working with Coalition Forces.

After a few minutes the ODA team came out with two detainees and some evidence. I don't know what they found but something that interested them was in a brief case and an ammo box. Apparently it was bomb-making material. Every SF soldier goes through language training as part of his training and these guys could interrogate any subject without the assistance of an interpreter but they had a few along for good measure. By 0730 we had secured the area and were headed out with two bad guys and the evidence that would hold them for a while. Not a shot was fired.

It was an interesting process taking someone as a detainee. As was often the case, the detainee was a male and the head of the household, so when we'd take him into custody, the wife, mother, and children would all be outside wailing for us to release him. My problem with that was that if the family members didn't want their relative taken or killed because of something stupid he did that seemed like a threat to coalition forces, they would have encouraged that relative not to do the things that would get him into that situation in the first place.

We then headed to the JCC to drop off some of our guys to relieve the previous group who had been standing guard duty for the last two days. Our mission ended when we made it back to Caldwell with no other incidents. All in all, it was a very successful mission. The New Year had begun on a good note and as a benefit we had no other missions that day.

After lunch an old buddy who'd been at medic training with me in Texas came by our CHU. Fog Head visited for about an hour and was happy to tell us that he was now assigned to do Preventative Medicine for the Regiment. His job was to go around to the different FOBs that the 278th was in and ensure that water was being treated properly and that food was being made and stored correctly. This was good for him because he had been assigned to Charlie Medical Company and wasn't a big fan of how they did things.

Charlie Med was the team that ran the Troop Medical Clinic. It was there that the majority of Regimental sick calls took place and front line injuries were treated. While it was a good place to get treated, historically, it was quite a bit political. From what Fog Head said, Charlie Med had an abundance of soldiers who were too ready and willing to stab another soldier in the back while jockeying for position, rank, or both.

That night I also got a chance to have dinner with CFB and Cankles. It was always good to see them because spending time with them gave me the opportunity to see this place from a different perspective other than from our platoon, the Double Deuce. CFB showed me some pictures of the areas he'd been to on his missions so far, one of which was Iran. He got to go into the mountains of Iran to encourage the Iraqi soldiers near that area to keep a better watch on those allowed to enter Iraq.

January 2

Today we had another patrol in the area where we found the bad guys yesterday. We found that when we gathered to get the Operation Order we didn't have enough vehicles to go out and accomplish our mission. We waited a few hours for the previous patrol to come back. Fortunately The Big Red One and Dark Wing allowed us to get chow.

The patrol still wasn't back when we returned. For the first few patrols while we were in Iraq this was the norm as we waited for logistics and organization to be completely worked out.

When they did return our orders had changed. Apparently some crazy Hajjis were supposed to be hitting our position with rockets at 1500 so we began our patrol in a different location. After arriving, we dismounted and began patrolling. I drove Big Red on this particular mission. Black Water 6 and his driver both dismounted, so when we had to move, the LT had to drive that vehicle. BW6 allowed his team to do that a lot. It wasn't the smartest thing to do and his guys got cocky because they were supply and office boys who had been allowed off post only because they were with BW6, the CO.

We drove around the area for a little while as the guys patrolled and checked out some houses. The usual routine we followed was to allow our dismounts (ground pounders) to get about 50 to 100 meters in front of us, a place where we still had eyes-on, and we would trail them along the route that they chose. They would enter a house and we would provide over watch for them. Should anything have gone badly, we would have been close enough to them to hear it live or get it over the radio and could have responded immediately.

Our dismounts led us into a huge field that was caked in mud. Driving around in it was like going four wheeling at home. That was something I found to be consistently true about the desert in Iraq. When it was dry, there was always the chance of a sandstorm. When it was wet, we couldn't go anywhere without getting caked in mud. It was fun to drive in but it was pure hell to walk. Boots took on a life of their own after a patrol and twenty layers of Iraqi desert mud would be attached when one returned to base. It sucked.

School was getting out for the day. Kids were everywhere. The second we stopped they swarmed us. Thirty or forty kids surrounded each vehicle asking for money and handouts. We tried to play nice American soldier for a while but there were so many of them we couldn't finish our foot patrol. Black Water Six got swamped when he started giving out Tootsie Rolls. Dark Wing, frustrated with the whole ordeal, finally made the call to stop the foot patrol and everyone mounted. We drove the back streets for a little while. As we did we came upon a group of men selling black market gasoline.

After clearing the gas area, we headed back to Caldwell. Some of us went to chow while others relaxed in the CHUs. About an hour after chow, Pretty and I headed to the gym to do some strength training. I had just gotten on the tread mill to loosen up when Tater came in and said some stuff had gone down in town and we had been called up to go and support.

Boo Boo offered his account of what happened on our quick reaction in support of one of our units taking fire. "I'm trying to think of what was going on. Doc was headed off to the gym and I was going to stay in my little CHU. Dark Wing came by in full gear and said we had to go support some other guys from Apache. My job when I got to the Bradley was to fill the radio up. When we got to the area of the shooting I realized I didn't have my head piece for my night vision.

"People were running around the place doing all sorts of things. Doc and I went to a corner of the building where a car was. I fucked up and didn't check the car but the next thing I know I was taking a leak and they were screaming for a radio man on the ground so I pinched it off and ran over and climbed up the captain's ass. I became Black Water Six Romeo for the next hour. I followed him around as we searched each area. I don't like Black Water Six. He pissed me off this afternoon on our patrol. He tried to take control from Big Red and Dark Wing. I don't think he knows as much as we do out there. He knows the people personally but we know the attitude of the people. He thinks the city is good and we think it's shitty. He wants to go into good places and we want to go into bad ones.

"We ran across this little tin roof thing of a bridge to where I was quickly acquired by the CO. That was pretty much the last time I saw Crowfoot and the rest of my team until we loaded back up. I got to tell Dark Wing that if anything happened in this house just to unload on it. As we were searching one particular house I looked in a car and found an AK-47. I tried to bring this to the CO's attention but he was preoccupied with searching the surrounding area. Finally he paid attention to me and upon further inspection we found more magazines than the family was allowed.

"Illumination rounds were fired. One particular time Doc was taking a leak when they went up and he was caught in the light with his wanker out. Black Water Six told everyone to mount up and we headed back out across another rickety bridge. I thought for sure I'd fall in. We loaded on the Bradley and as we were turning around to head back to Caldwell we realized that Mickey D had not put enough tension in the track. It popped a few times while we turned. Everyone in the back agreed that the track was going to fall off.

"We headed back to Caldwell. The wind was blowing like it was trying to pay back the insurgency for doing their dirty little deeds. Doc went to take a shower and what he didn't know was that while he was away earlier I had made a Fee-Fee and put it under my pillow. [A Fee-Fee was a sock filled with lubricant that was used to assist a guy with masturbation.] When he left for the showers I whipped it out and did my business."

Throughout our stay in Balad Ruz there were always gunshots being fired in the city. Ninety-nine percent of the shots were fired in celebration of something. Very seldom were they fired directly at our position while we were at the mayor's office. The first few times we stood watch down there though was nerve wracking because it was something we just weren't used to in America. After a month or so of standing watch though, it just became routine. Unless we heard a zip, crack, or ding, we didn't think about the gunshots much.

JANUARY 4

Yesterday we did a patrol in the market area of Balad Ruz. It was good to be out in an area where a lot of people were. Granted, we had to stay much more alert because our senses were being bombarded with all sorts of input – sounds, sights, smells - but we did the patrol very smoothly. We did hear some gunshots south of our position on the main street. They sounded as if they were a few blocks from us. Why we didn't take action in that direction I'll never know. Every time we had an Op Order we were told that we would engage any direct or indirect gunfire. By that I thought that we would actually investigate anything, anywhere, anytime. That wasn't the case today. After coming to the first traffic circle we loaded back up and drove down some small back roads.

One of our comrades, who had bent sent home to nurse a knee injury while we were in Shelby, was waiting for us when we returned. Gunro had just come in that day and was on his way to join the rest of our platoon at Bernstein.

Later that night Boo Boo, Pretty, and I settled down and watched *The Patriot.* Word got to us that the Iraqi Police and Iraqi National Guard were in a gunfight with each other in downtown Balad Ruz. Our B Team, the second of two fire teams in our squad, was headed out to patrol while this was going down. After their patrol was done we asked them about it and they said they saw nothing but did pursue some suspicious characters in a blue Opal. For the year that we were in Iraq, we chased the phantom Opal. Anytime a suspicious vehicle was reported, it always seemed to be an Opal and it always seemed to elude us.

Today we did another morning patrol on the other side of the market. The market was an interesting area. There wasn't anything there that I really wanted or needed to buy but somewhere in all the hustle and bustle there was a bread maker. The Iraqi bread was some of the best I'd ever had. A fresh vegetable market seemed to be the biggest attraction. All the clothing, shoe, and household goods stores were strung throughout the city on the side streets. The farmers left home for the market some time between 0430 and 0500 and stayed there for about twelve hours. The people swarmed the market area from about 0900 until about 1600. The traffic was always horrendous during market hours. Traffic cops lined the streets but didn't really do a damn thing to help the flow of vehicles move at a good pace.

The guys who had dismounted went into one or two houses and were headed back to our position when they saw a blue Opal. They quickly ran to where it was and stopped the driver. Getting everyone out, they searched the vehicle, finding an unauthorized AK-47. Our guys seized the weapon. CFB had his picture taken with it. Joe always seemed to find a good opportunity to get a sweet "war picture" made so he could send it home and impress his friends. The stop turned out to be nothing more than a sheik who was being driven to a meeting. His driver simply hadn't registered his firearm properly. We still took the AK and encouraged the driver, for the sheik's protection, to get his paper work together.

Other than that, all we did was drop off and pick up the CO at the JCC. The rest of the patrol went without incident. For this patrol I manned the 50 caliber machine gun on Dark Wing's vehicle. As a medic I was not supposed to be in an aggressive firing position. I was not to shoot unless that was my last recourse or I was protecting a casualty or defending another human being. But since I had come to the unit under the guise of an infantryman first, medic second, I was afforded the opportunity to play around on the big guns every once in a while.

****

From: <Malcolm>
To: <Mom>
Sent: Tuesday, January 04, 2005

You worry too much, Mom. Yes, I'm still a medic but since our reattachment my primary service is as an infantryman with our company. With our squad I'm still "Doc." It's all good. I take my bag every time we go out. But don't worry, there's another medic around should anything happen to me. Just kidding.

JANUARY 9

Today I saw first hand why Boo Boo was still a Private First Class – his attitude. After coming back from two days of eight hour watch rotations at the JCC everyone was tired. We all were a bit crankier than normal because we were told we would be going back to take care of a downed Hummer that should have been taken care of as the relief platoon was turning over with us. Regardless of how we felt that was just part of the duty we pulled. We always had to be ready to deal with Mr. Murphy and we always had to be ready to improvise, adapt, and overcome.

As everyone was unloading his gear I was on the Hummer repositioning the Mark 19 grenade gun that I had taken down prior to hearing we were going back out. No one came back to relieve me so I could take my gear back to the CHU. Fortunately Spider was still around, so he and I grabbed our gear and left the Hummers in the hands of another soldier from first platoon.

When I got to our CHU I found the door open and Boo Boo outside just hanging around. I took my gear inside. All his gear was in the doorway. That would not have been so bad except that the trash and brooms were in the door way too. I also noticed a microwave on Boo Boo's table. He asked me if I wanted to fix and keep it. I told him I had no use for it. He then asked me if I would throw it out and my response was, "I'll throw it out if you get all your shit out of the way. It's everywhere."

Boo Boo had been used to getting everything his way. I wasn't going to play his game. He soon found out what it was going to be like to live with another person who didn't think he hung the moon. He was mad and proceeded to throw all his gear at the wall where his bed was. He threw the microwave out the door as Nasty was passing by, almost hitting him in the process. He then stormed out of the CHU and I didn't see him again until after I returned from lunch.

Granted, everyone has attitude at some time but that kind of reaction was ridiculous. It was obvious to me why he'd been demoted twice. The shame of it was that he took pride in being busted. He thought he was telling people how tough he was.

The whole "bad seed attitude" was something I didn't really understand about this particular squad. It was like the more useless trouble in which they could get themselves the prouder they were of the negative results. They often bragged to each other about how cool it was that they had gotten busted and were still allowed to be in the Guard. They were always trying to do some kind of stupid shit – self-destructive to themselves and those around them. Dark Wing led the way with his stories of incompetence and they followed him around like a group of ignorant, little ducklings.

We've been at the JCC. Bong Water Blue was our Non-Commissioned Officer in Charge (NCOIC) but one wouldn't have known it because Crowfoot, Crazy, and Pretty didn't allow him to run the detail. They were Dark Wing's boys so they ran over senior ranking Sergeants because they knew Dark Wing would back them even if they were in the wrong. It was leadership situations like that that made me question the validity of our chain of command. If one kissed the right ass one could do anything in our squad. It was much like life. If one didn't, one was relegated to menial tasks no matter the rank. Again, it was much like life. The value of the work proved nothing when one was involved in a "good old boy" system. Only the agenda of the person who gets his ass kissed was the priority.

Nothing much happened at the JCC for the two days we were there. There was the routine random firing of weapons in the surrounding areas but nothing to get worked up about. We simply reported the shots from the positions on the roof, to the radio room, and then logged it so that if anything more happened we could refer to the chronological timeline of what had gone down.

Some of my IA buddies - they kicked ass!

January 11

Last night's adventures took us on a new twist, one that we'd all been prepping for since Shelby but one that none had expected to see this soon. As I was packing the contents of my aid bag to protect them from the weather, a call came out that our squad was to load up an hour early to respond to a suspected IED. We rallied relatively quickly considering that some of our guys had already headed for evening chow. The Op Order was given and the guys loaded into their assigned Hummers and Bradley.

The back of the Bradley was packed with Pretty, Crowfoot, Tater, Boo Boo, Mickey D, and me. For those who have never had the opportunity to ride in the back of a Bradley, just pretend you're sitting in a three and a half foot room that jerks and rattles when it moves, is cramped, has no good air, is louder than a Kiss concert, and you are the target of both improvised explosive devices and rocket propelled grenades. Five people were tight but six soldiers in full battle rattle were like trying to climb back into the birth canal. Not that it couldn't happen; it was just bit tight.

We were going into Balad Ruz to check a suspected IED that the ING had found along the median of Main Street. We reached the area with the expectation that we were going to see some real explosives. We also had to be on alert that the insurgency might have placed a decoy so that they could check out our tactical procedures and react to them accordingly in the future. At the very worst, our battle plan took into consideration being lured into an ambush.

Because the IED was reported by both the local Iraqi Police and ING, there was some confusion as to its exact location. We spent the better part of 45 minutes searching the area for it. I was later told that we drove past the IED a few times. That wasn't the smartest thing to do, but in our position in the back of the Bradley we had no real clue as to what was going on outside.

Clementine, the Bradley's gunner, was the guy who eventually found the suspected IED by using the vehicle's thermal imaging and night vision systems. Bong Water ordered the Bradley back a few hundred yards. The other vehicles with us pulled into their security positions down other streets to cordon off the area. Dark Wing and Crazy were in a Hummer closest to the IED. At one point during the mission, they walked up to it to see if it was, in fact, an IED. That wasn't the protocol for dealing with IEDs but after passing by it numerous times and then having been told we were waiting on the Explosive Ordinance Disposal team, they got restless.

Finally, the dismounts were allowed to get out of the Bradley. We set up a security perimeter in our section, still unaware of the location of the IED. We could have been twenty feet from it and nobody was telling us anything so we simply stood our watch for the next hour. Eventually EOD arrived with their little robots and flash suits. Big Red ordered us all to mount up, so for the next few hours, those of us in the back of the Bradley had no clue what was going on. We sat in the back and tried to figure out the most comfortable positions in which to sit. It was cold and raining that night. We were miserable in the back and all of us needed to smoke, spit, fart, and piss.

The conversation in the back of the Bradley was not very eloquent or politically correct but it was quite interesting. Everything from past sexual relationships to racial views was talked about. Pretty led the conversation with a question about the effects of gravity at the center of the earth. His question was something like this: if a man in the U.S. dug a hole to the center of the earth, which way would be up? His thought pattern was that if that man stood on the surface of the earth and he searched the heavens, he was looking up. If a man in China did the same, he too would be looking up. If they were able to tunnel to the middle of the earth, when they got there, would they be facing up or would the pull of gravity make them weightless? Needless to say we all laughed at him but it led to a lot more questions and conversations.

By the end of our random conversations, most of us believed that Pretty was a racist and Boo Boo was really a white man trapped in a black man's body. In reality, Pretty just hated the worst acting people of each race, the ignorant assholes who make everybody look bad. He hated niggers, white trash, spics, and anyone else that made humanity look like trash. It was funny watching and listening to him and Boo Boo go back and forth on why one should or shouldn't like a certain person or race. He had Boo Boo worked up so that he was cussing him every other breath.

One thing that Pretty was not afraid to talk about was his sexual exploits with women. He ranked himself right up there with Wilt Chamberlain and Gene Simmons. He believed he was a man among men. If he was attracted to a woman he pursued her until he got what he was after. His stories were usually pretty funny too.

A large part of the time was spent discussing which drugs each of us had done in our lifetimes. I've done a few in my past but my particular drug of choice was alcohol. I used it as a crutch to interact with people. I guess that most people do who drink. I quit drinking in May of 1996 and didn't touch it again until I left for Camp Shelby. I don't know exactly why I wanted to drink again other than I thought it was a good crutch to interact with some of the people I knew.

I knew that I ran the risk of going back to where I was with it - out of control. I prayed that God would protect me because I did know where I came from with it. I knew how out of control I had been and I was scared as hell to go back there. I didn't want to drink simply for the sake of getting drunk. That was stupid.

One of my papers in seminary was about the biblical view of drinking. From what I studied I could not prove beyond a shadow of my doubt that drinking was either pro or con in the Bible. The Word of God has both views in it. It tells horrible stories and proverbs about the abuse of alcohol. Noah was a good example of what getting drunk can do to a family. But throughout the Bible there were times when wine was consumed without issue. Jesus' first miracle was turning water into wine – the best at the wedding ceremony as a matter of fact. Temperance seemed to be the issue God wanted and continues to want to make, not only in alcohol but in all things. That was what I shared with those guys that night in the back of the Bradley.

We had been on the mission for over four hours and in the back of the Bradley for the last two hours when we got the word that the EOD guys were going to blow the IED. They gave the word that they were going to blast, did their countdown, but the first blast was only as big as the charge. They didn't think they had blown everything. They were going to have to blow it again.

We were allowed to dismount to stretch our legs and do our business. While we were out we asked if we could observe the next blast since they had to do another one to ensure that the "bomb" was blown and the area was safe. Big Red allowed us to observe. He had us move father back from the blast area and wait. Thankfully the EOD guys set the charge up pretty quickly and detonated the bomb. It was loud and the concussion waves hit us at a couple hundred meters. I thought there would be a bigger fireball from the flash but there wasn't, just a good sized blast and some smoke.

After the blast, EOD cleared the area. They gave the word and we went to check out the site and take pictures for the intelligence report. Once done we mounted back up and headed to Caldwell. The ride was uneventful and we returned with enough time to download our gear and make midnight rations (midrats). I came back from midrats and watched *The Last of the Mohicans* before I went to sleep. The whole mission lasted almost seven hours. Five and a half of it was spent in the back of a Bradley. That sucked.

What an IED can do to a Hummer

****

From: *<Mom>*
To: *<Malcolm>*
Date: *Sun, 2 Jan 2005 05:39:25*

Hi there! What do you do on a raid? Do you go into homes? I thought things were supposed to be pretty peaceful and friendly where you are. Please be careful!!!
Have a good day!

****

From: <Malcolm>
To: <Mom>
Sent: Wednesday, January 12, 2005

Things are pretty peaceful where we are (right). Yes, a raid is when we go in homes or into buildings without warning the people we're coming in. Sometimes it has to be done to catch them by surprise. We don't do a lot of it though. What's going on today back home?

JANUARY 14

Today's mission was to clear the western boundary. The mission began with a 0300 wake up. We were at the Hummers by 0400. Our actual start time was 0500. During the hour between when we showed up and when we actually left, we did last minute equipment and personnel checks. We also got our Op Order. This was a routine thing before every mission. On this particular patrol I would be riding left back passenger. Tater drove and Pretty was the gunner on the SAW. Dark Wing acted as our vehicle commander. Crowfoot rounded us out as our final passenger.

The run out to the western boundary was uneventful except for fog and long gas lines that created a bit of traffic for us. It was amazing to me that a country so rich with oil had the problems it had with getting gas to its people. Without a doubt the conflict here had taken its toll on the infrastructure of Iraq, but it seemed to me that something better for the Iraqi people would have already been in place. I suspect Saddam's power was a lot deeper than I could or would ever conceive.

After we passed the western boundary, we turned around into an open area south of Route T1 where we dismounted and fired our weapons. To that point, we hadn't seen any conflict like in the areas surrounding Baghdad or Fallujah. The Marines were in those cities kicking ass. We wanted to be there too but we just didn't have the same mission. We had our AO and fortunately for us, it wasn't very active at that moment.

We shot at rocks and bottles – lame targets in contrast to what we could have used those rounds for. Thankfully no one was returning fire and we were not receiving any casualties. Sometimes I felt like the glamour or romance of war overshadowed the deep hurt that a soldier really does experience.

Throughout history, soldiers, including all combatants, have seen death firsthand. Many have seen it in the most brutal ways any human could ever imagine. They have killed, watched fallen comrades writhe in agony and die, and to some degree, the fortunate ones have experienced the twilight of death themselves. I've often wondered how some of the overly vocal guys would react under intense fire. That would be a different story, one that our guys, for the most part, had not been used to.

After expending a few rounds we headed back into Balad Ruz. The city was alive by now. People were everywhere and the traffic was horrid, again due to the gas lines. We went to a particular section of town that looked like a place where chop shops ran rampant. We dismounted and began to patrol the area with Dark Wing as our team commander and Shrek on point. We patrolled the area for a while, finally ending when we found a black market gas seller. The guy was questioned and then picked up by the Iraqi Police. I don't know what happened to him but I suspect he and the IP worked something out with the IP getting the best end of the deal.

JANUARY 15

The 0200 patrol was slotted as a mounted patrol. We would only dismount if we encountered anyone out past the 2300 curfew. As fate would have it, we encountered a vehicle out past curfew when we made our first turn down a street in Balad Ruz. When the dismounts were in the back of a Bradley we had no real way of knowing what the situation was outside unless we were being told by the driver, gunner, or Bradley commander. We got none of that as we pulled up to the vehicle. We were only told that a car with its lights on was in front of us. One of the key things in any situation was to know the surroundings.

As we dismounted I realized we were in a familiar place. Pretty and Crowfoot had already turned the corner and were on their way to the front of the Bradley by the time I figured out exactly where we were. We were in front of the hospital. Pretty and Crowfoot had the driver slammed up against his car with a shotgun an inch from his forehead. They were screaming at him, asking him why he was out. Since neither party could speak the other's language, neither could convey their confusion as to what was happening. All that could be seen was that this guy had his hands up trying to remain as calm as possible while two soldiers, doing what they thought was right, scared the absolute shit out of him.

When I got to the car I noticed a small boy in the front seat. He looked a little sickly. I yelled at the guys to ease off the man and tried to get them to understand that we were in front of the hospital and the guy was probably taking his kid to get treatment. The situation continued to escalate as Pretty, Crowfoot, and Boo Boo put zip tie cuffs on him.

I continued to try to get them to understand that we were at the hospital. They finally caught on when three other men came out of the building with a child who had an IV in his arm. One of the men was older and seemed to know the man who was now in cuffs. Tears filled his eyes as he tried to tell us what was going on. You could tell that he was frustrated with the situation. He just wanted to take care of his kid and his friends. I tried to reassure him and let him know that it was a misunderstanding. I hated than none of us understood Arabic.

After five minutes or so of failed attempts to communicate, our guys tried to take off the man's cuffs. The older gentleman was still upset. The first man's son, who was still in the car, looked completely frightened. In the scuffle, the cuffed man had cut open his toe. I tried to tell him that he should go into the hospital to have it looked at. It was to no avail and I couldn't treat him because we were moving out. Eventually everyone got calmed down enough so we got the man's name and tried to convey to them that they should take the matter up with the mayor's office. That was the only recourse for actions that we had done that may have caused someone or his property injury.

From my perspective the whole incident was a poor example of a military execution and leadership. The guys who could see where we were should have been telling us our location prior to our dismounting. Our commander should have been on that immediately. It's called situational awareness. Just from a simple head's up like that, we could have approached the situation completely differently.

Crowfoot and Pretty should never have gotten on the guy like they did. As a soldier, and particularly a US soldier, we walk a fine line that sometimes borders the powers of God. We can choose to take another man's life. We can choose to abuse our power and intimidate those who don't have equal arms. In the first place, we were all being covered by a Bradley Fighting Vehicle. If the guy had done something stupid he would have been blown to bits before any of us got to him. Secondly, one of the first two on the scene should have taken the guy away from his car and the other should have inspected the car for anything out of the ordinary. Then they would have been able to see that a child was in there and that the car was safe. Instead, adrenaline and arrogance took over.

We should have searched all the players as well if we were so concerned about these people being out past curfew. Not one time during the entire evolution did I see a Joe search any of the men. I include myself in that accusation. I also should have been more assertive in checking the attitudes of our soldiers. I wasn't opposed to showing or using force but there was a time and place for it. This had been a bad habit of our team since day one.

Dark Wing has been really good at motivating the guys, but he and Crowfoot have been horrible at keeping them in check. Every mission started out with a string of taunts towards the Iraqi people about how the Deuce was going to shoot up everything and kill everyone that looked at the squad crossly. I never took part in that. It wasn't professional. It wasn't who I was or who I wanted to be. It was redneck and rednecks embraced that attitude.

When we got mounted again, Crowfoot and Pretty talked about the situation like they had done the best thing ever, but after a few minutes Pretty began to realize that he had used excessive force and shouldn't have been such a bully to the guy in front of his kid. I was saddened by the whole event. No wonder other nations want to kill Americans. We have a tendency to be very arrogant and when we have guns in our hands even more so. I'm all for having a gun, especially in a hostile environment, but having one doesn't make you a badass. It just makes you a guy with a gun. If anything, one should be more careful with his emotions when he's got a gun in his hands. The situation could have turned out much uglier.

We continued on our patrol after leaving the hospital area. In our leader's infinite wisdom we headed down a street that should only have been maneuvered by dismounts. Needless to say, we ran into a situation. Our lead Bradley got stuck in a shit ditch. We spent the next hour or so attempting to pull it out with the other Bradley. Finally we were successful and managed not to have any more incidents that night.

I will say this: there were shit pools and shit ditches all over that city. All the sewage ran out into the streets. Trash was everywhere. That was the norm. I had never seen anything quite like it. At that time, I'd been to over twenty countries but none of them had been in such a bad state.

January 17

Yesterday's mission was a trip into extended boredom. Our mission began with our being at the Bradleys an hour and ten minutes prior to SP. We showed up an hour early to ensure that everyone was accounted for and everything worked properly. The Deuce was in the habit of showing up ten minutes prior to that hour so that we could look like we had our act together. I suspected a lot of troops did that in hopes that they would stand out in front of the other men in their units. Then there were the units who didn't really give a flip about impressing anyone so they would just do their job and do it right the first time – no need for accolades.

After getting our Op Order we loaded up and headed west on Route T1. At Tactical Check Point (TCP) 4 we turned south on Route C1. As stated before, riding in the back of a Bradley was pretty uneventful because we did not get a full view of what was happening outside the shell of the vehicle. There were a few portholes from which to look out, but they gave us a limited perspective of what was really going on. The noise inside an idling Bradley was bad, but it was made even worse as soon as the vehicle was in gear and moving. I would always wear hearing protection.

We rode for about 40 minutes and finally stopped at CP 12. Next, we dismounted from our Bradley and pulled security around the vehicles. To our east was a housing compound where we decided to do a soft knock and search. We found nothing there except a bunch of black phantom women and a few men and kids. I felt bad about walking into their home because they were in the process of cleaning the floors. I knew we had to do it though, so Nasty, Little D, Jonesy, and I searched it as quickly as possible.

The search turned up nothing so we returned to the Bradley and went back to the convoy area where we again dismounted and pulled security. Our mission at that time was to be alert for any suspicious people, check any cars we wanted to, and provide security for the retransmission team while the rest of the convoy went further south to an area where they would recon for a future raid. The retrans team was a mobile communication center that retransmitted radio communications between our units that were out of normal radio range back to either the JCC in Balad Ruz or headquarters on Caldwell.

One of the more interesting things we did run into was a group of falconers. Two jeeps came through our position, each with two falcons. They were awesome birds. The gentlemen who owned the birds seemed more relaxed and a bit more educated than all the other Iraqis we had interacted with that day. They had found something that they enjoyed doing with other guys who were like minded and apparently were good with their birds. The second car of falconers had another bird on the floor board. The falconer loosely told us that the bird was for his falcons. He also had a shotgun below the passenger's seat.

Our Standard Operating Procedure (SOP) was to check for registration of any weapons that we found. If the person did not have registration for the specific weapon, and he was only allowed one, then we took the weapon, cataloged it, and turned it over to the IP. I didn't do it to this particular guy. I'm an American who believes in our right to bear arms. Besides, he had his falcons with him and didn't appear hostile towards any of us. I didn't take his gun. I simply told him as best I could that he needed to have registration for it, took a few pictures of his birds, and then sent him on his way.

The highlight of our time at CP 12 was when Bong Water relieved himself in the ditch in the middle of the split in the road. Nasty, Little D, and I were scanning our area when out of the back of the Bradley ran Bong Water. He said, "I gotta shit," and ran into the ditch with toilet paper in hand. I had to give it to him, he didn't care that we were smack dab in the middle of an intersection or that kids were walking past our position all the time. He just dropped his pants and began his duty. Being the ever ready photographers that we were, Nasty, Little D, and I started taking pictures of the whole thing. It was hilarious – Bong Water taking a dump and the three of us laughing at him and taking pictures. We shared them with all the other guys as soon as they returned from their part of the operation.

After their return, we regrouped and headed back to Caldwell. The 40 minute ride was uneventful and most guys took naps. In the other Bradley, Boo Boo fell asleep and Mickey D and Dark Wing smeared peanut butter on his glasses. He awoke to find that he couldn't see clearly and was more than a little ticked off. He retaliated by spewing water and MRE shake power on everybody in the back.

January 18

Today we worked on the Bradleys or should I say, some guys worked on the Bradleys. I sat beside one for two of the most wasted hours I've had since being here. Finally I got fed up with the stupidity of being out there and got the go-ahead to return to my CHU and clean my weapon. I quickly finished that and then went to the computer center to check emails.

Today's batch of emails included one from a friend whom I had not talked to in years. I was very fortunate in that there were some folks who routinely sent me emails. I really enjoyed hearing from them, but the ones that were my favorites were the ones from people whom I hadn't talked to in a long time. They would write me out of the blue and make my day. The war had given me the opportunity to reconnect with some old mates I was sorry to say that I hadn't been keeping in touch with over the years. I suspected that was part of life but I thanked God that He saw fit to reunite us. It was definitely a good thing.

JANUARY 19

We were slotted for an early morning wake up, patrol, and then JCC duty. We headed into town and were dropped off in the palm grove area. The Bradleys continued on a mission of their own. They would pick us up when we were done with our dismounted patrol which turned out to be pretty cool.

We dismounted near a block in Balad Ruz where we had suspected VBIEDs to be made. After securing the immediate area, we crossed over a wall into the palm grove. It always amazed me how beautiful the palm trees were especially at night, when we couldn't see the rest of the city or the filth that surrounded them. If this particular area were in any other part of the civilized world, the area would be a resort. But in Iraq it was just a beautiful oasis in the middle of a bunch of ugliness.

Apparently, in the area we were stationed, there had been huge plots of land with both palm and date trees. From what we were told by the locals, Saddam's Army had come through and leveled the majority of them for no other reason than he was and still is a despot. Fortunately there were a few groves left and we were able to visualize, to some degree, what a beautiful place the area could have been. Maybe it really was where the Garden of Eden and Babylon had been.

Patrolling through the groves took a long time and we only went through a small portion, south of Balad Ruz. Pretty walked point and throughout the patrol I felt like I was in some exotic country like Panama or Vietnam. Obviously the vegetation in Iraq was nowhere near the density of those jungles, but at dawn it did have that spooky kind of feel like I had read in history books and seen in movies.

We came upon a building about 45 minutes into the patrol. The front gate to the compound was locked so Mickey D charged the door and hit it hard enough for us to enter. The dead bolt bent and in we marched. The main building was like most buildings we had been in thus far, mud and sticks.

Inside we found a scuttled car that had whiskey bottles strewn about it. Further in the building we found sacks of dates. We assumed it to be a date storage building. We poked around for a bit, finding concertina wire and some palm branch cutting knives. They were more like sickles. We took them. They would prove useful in further probing operations. Nothing else was there so we took a few hoo-ah pictures and then left.

Pretty led us to an area where he thought we could cross over into a field but found that when he crossed the wall, there was a canal that couldn't be seen from our side. It was too deep and wide for us to cross without getting dysentery so we doubled back to the building and found a way to get across the wall at that location. A family from a house across the canal watched and laughed as we clumsily cleared the wall with all our gear and we were barked at by the local contingency of dogs.

The field that we had to cross next was large and too open. The only thing we could have used as concealment if someone had wanted to fire at us was the tall grass. Not very safe from gunfire for sure, but I guess it would have been fine should the need have arisen. More than likely they would have only seen us if we had gotten up to shoot or advance. Low crawling would have been a pain but we could have done it without disturbing the grass for the most part. We moved across the field without any incident, making a few stops along the way for the occasional shots of the patrol.

After crossing the field we headed back into the southern part of town. Near the first house as we reentered the palm grove, we spotted some desert foxes. No, they were not hot Iraqi women. They were desert canines. They ran in front of our patrol and after two went by we were given the go ahead to shoot the next one that showed his face. None did.

As we entered the more inhabited area of the grove, we ran into another falconer. His falcon was big and beautiful. I suspected that he and his friend were just getting the bird out for the morning air. The owner allowed us to take a few pictures and then we moved on to link back up with the Bradleys.

We remounted and headed to the JCC. At the JCC we put our gear on our racks, relieved the previous watch group, and assumed our positions on the roof and in the radio room. The first day at the JCC has always been the day for me to catch up on sleep that I've missed. When I wasn't standing a watch I was in the rack, knocked out. It was good. It was the only thing I looked forward to when we got to the JCC because it was a nasty place and all the people there wanted something from the Americans. If we slept through it we didn't have to deal with all the other bullshit. That was nice.

January 20

Apparently today was a Muslim holiday around the world. I wasn't completely sure which one it was but the Iraqi national news was televising the religious celebrations all day long. We saw some of the celebrations from the JCC as well but what we saw came with a little gunfire. It seemed that every time somebody in Balad Ruz wanted to celebrate there was a rain of gunfire from wherever they were partying.

This type of watch went on all day; so as one would expect, it was less than exciting. Our biggest adventure was trying to find a local runner who would go down to the market and get us bread, chicken, and kabobs. This proved to be a bit harder than our previous times at the JCC because the guy who usually did it wasn't around. Finally, we did find some boys to do the run for us so we were able to enjoy some of the finer things of Iraqi culture.

A few of the guys scored some liquor and pills today. They weren't discreet about it in the least. Sometime during the day Mickey D took some of the pills. They were Parkinson's pills and were supposed to give him a little bit of a high. That wasn't the case. Later on that night his body paid for taking them. He got sick with fever and nausea.

Worm had also scored some Valiums. He and Percolator popped a few while they were on watch. Drugs were readily available on the streets and from the IP. Guys used them to forget about where they were – deal with whatever they were dealing with. I usually didn't have a problem with guys popping pills or drinking. I just thought taking the pills was stupid. Taking them while on watch or on patrol was unforgivable because that put other soldiers' lives in danger. Who was to know how a person would react under intense circumstances while he was baked out of his mind? I didn't want to find out. Neither did any of the other more responsible guys.

Pretty got sick that night as well. I don't know that he took any of the pills but he had it coming out both ends. He'd had issues down at the JCC before where he'd shat his drawers, but this was a bit more because he was throwing up. It was a bad night for both of those guys.

January 21

Our team was awakened for the 0400 to 0800 watch and both Mickey D and Pretty had to be relieved of their watches and come down to the TOC because they felt so bad. I didn't have my med bag with me and nobody had a Combat Life Saver (CLS) bag, but fortunately the Blue element was patrolling the town and CFB was with them. He was able to come by and hook me up with med supplies. We snagged some IV bags and needles from him so I could introduce liquids to the guys.

I was a good stick usually, but the first stick I did to Pretty didn't go because the catheter on the needle wouldn't penetrate his skin. I had to stick him a second time. It went in fine. I was good at my sticks but I hadn't quite mastered the art of a small blood trickle between the stick and the application of the tape and tubing of the IV. The missed stick left me with no needles with which to stick Mickey D and introduce fluids to him. It wouldn't have mattered anyway because Pretty absorbed the two bags of fluid. I gave Mickey D some ibuprofen for his headache. I knew why he was sick and had little compassion for him. He had just wanted to get high. His careless attitude caused another man to have to stand his watch and put the rest of us in danger because we couldn't depend on him. That became a routine thing for him while he was with us.

After Blue was done with their patrol they came back to the JCC and picked up Pretty and Mickey D. They were kind enough to leave behind two replacements for us. It was good for us but bad for them because we ended up staying at the JCC longer than our normal time due to the amount of IEDs reported after 1030.

Six were reported in town that day. EOD was called out and did their thing where and when they could. The worst issue about calling out EOD was the amount of time it took for them to respond. Even when the IED was only a mile down the road, they usually took two to four hours to get off base and link up with whichever unit was sitting on the bomb. There was a process they had to go through to identify the type and size of the bomb before they ever got to the scene. Most times it was very frustrating for those of us who had to wait. We wanted the thing blown so we could continue our mission but we had to wait on them to do their deal first.

The Bradley crews got smart after a few run-ins with roadside bombs. The crews would stand off at a good distance and then fire the 25 mm rounds on the ones they came across. Not only did this course of action speed things up but it also allowed EOD to focus on what they were doing and not be rushed from one site to another. We definitely didn't want them to make a mistake. The frustration of sitting on a probable IED, having it blown, and then getting the word that it wasn't an IED at all got to be too much sometimes.

At 1630 we finally got relieved. We headed back to Caldwell only to be stopped along the way because of a suspected IED planted in a dead dog. As it turned out, the dog had been hit by one of our vehicles earlier in the day and the impact of the collision caused the collar of the dog to move into a position that looked like wires were coming out of its mouth. Anytime we saw something like that we got suspicious. It was a good thing – kept us on our toes.

That night Mickey D got drunk on some of the whiskey that he'd gotten in town. He missed the next mission because of a hangover that he played off as sickness from the JCC. Dark Wing knew it and covered for him. They continued to piss me off. Again, I could handle guys doing drugs and drinking on their own time; but when that became an issue that other people had to deal with, it became a problem that needed to be eliminated as quickly as possible. I never saw that happening with him because he was one of Dark Wing's boys. He could do no wrong in Dark Wing's eyes.

JANUARY 23

Yesterday was the day of days. It was our capstone mission. We were slotted to siege an area south of Caldwell, rounding up insurgents. We would also confiscate weapons caches. This was the charge of Apache. We were the shit - the 278th gift from God! We were all told that it was going to be an historic event. The only thing it ended up being was a big joke and an even bigger waste of time.

The day began just like all of our morning missions: up at 0400 and down at the motor pool no later than 0450. After ensuring that all our last-minute checks were good, we were given the final Op Order by the CO. We were told to expect rain later in the afternoon. I looked at the sky and only saw cloud coverage; it was coming sooner than he thought. We mounted our vehicles just as ODA and the ING pulled in to join us. With our team loaded on our Bradley, an attempt was made to raise the ramp. This was the beginning of the fiasco. Mr. Murphy was out and about.

Somehow our hydraulics had leaked in the leverage unit of the Bradley and the ramp would not rise back to position. Fortunately our crew was quick to respond. Old Blue, Clementine, and Priest worked feverishly to put new fluid in and had the ramp semi-operational within a few minutes. LT and the CO were both at the back of our Bradley discussing the situation. I made the comment that Mr. Murphy was going to be with us all day.

After a few failed attempts to raise the ramp and a long string of cursing, our situation finally got squared away and we began our convoy off the base. Things seemed to be going smoothly ... for a while. We made it down Route T1 without any incident and made a left at TCP 1, turning south down Route C1. Our plan was to meet up with the Retrans Team at CP 12, refuel the tanks, and then move to our assault position further south.

We arrived at CP 12 with a drizzle falling and set up security across the canal from the Retrans/refuel site. The tanks pulled in to refuel. This was actually when the mission started going down hill at a breakneck pace. The mission was scheduled for the previous week but was cancelled due to heavy rains and mud. The CO made the right call in that he believed the environment posed too many hazards for the vehicles. I kept wondering why he didn't make that call again this time.

While we were stopped at CP 12 it began to rain hard. We were given the word to start convoy again. No sooner had we begun to roll then we got the word from Rugger that our number two tank was disabled because of a fuel line leak. The engine was shot. The CO held the convoy up as the third tank towed the second tank back to the refueling area. Its mission was over.

As the tankers were getting everything hooked up, a vehicle coming toward our convoy decided to take another route to avoid us but got stuck in the mud in the middle of the field it had tried to go through. Big Red, thinking that this driver's actions were a little suspicious, suggested to the CO that his vehicle and one more should go to the vehicle, search it, and offer assistance to the person if everything was in order.

The CO gave the go-ahead. Big Red and his element proceeded to the area. Once there, they saw for themselves that the route taken by the Hajji was not a wise one for our vehicles to traverse so they began a foot patrol to the vehicle. This extended the time we had already planned to be at CP 12. We were supposed to assault the area no later than 0800 but it was almost 0930 by the time we started our convoy again.

With all operable vehicles back in line we resumed the convoy. We rolled about a quarter of a mile when the word came over our radio that one of the 113s had gone down and couldn't finish the mission. The CO called a quick stop and we waited again as another 113 towed the downed vehicle back to CP 12. Troops were shifted about to ensure all infantrymen had a ride to the objective. The convoy began to roll again.

We made it to CP 22 with no other malfunctions. The 113 that had been the tow vehicle did have overheating warnings but was able to continue. At CP 22 the track vehicles split from the wheel vehicles. We headed east around a town that our recon the week before had told us would not be a wise passage for larger vehicles. The Hummers proceeded through it. The tracks turned south and began the final leg of our movement.

After a hundred meters our Bradley crew was told by the lead tank that we were dragging barbed wire. We immediately stopped movement and I dismounted to see what damage had been done. The wire was wrapped in the right track and axels. I tried to pull the wire out by handling it as the Bradley rolled in reverse. This took a lot of the wire out but I needed more help to ensure that the Bradley didn't roll over the slack with the other track so I had the other dismounts get out and assist me. Nasty pulled security as Little D and Boo Boo helped clear the way for the vehicle to move in reverse.

We spent about five minutes working on the track, pulling and cutting wire. After everything that could be removed was removed, we remounted and proceeded in convoy again. The weight of the tracks and wheels cut some of the wire off but a lot of it was still wrapped around the axels. Fortunately it wasn't bad enough to slow us down. We continued south to the objective.

Our track and wheel convoys hooked back up in a field south of the second town we had bypassed. We continued movement south, down the canal system. After some time we had to stop again. Bong Water and I dismounted to see if we could get any more wire removed from the track. We were able to remove some more but not all of it. We remounted and continued mission.

By this time the rain had really picked up. The fields we were crossing looked like swamps. Water was everywhere and we were kicking up mud like we were riding around in Big Foot. It caked the tracks, the vehicles, and our boots every time we moved outside the vehicle. The inside of the Bradley looked like a mud rug had been laid on its floor. We had to be careful when we moved to get out of the vehicle because the mud on the floor would stick to the mud on our boots like cement if left in one position for too long. If we weren't careful when we went through the hatch, wham, our dicks would be in the dirt. Actually, it would be our faces in the mud.

Eventually we made it to the objective. Our particular role in the assault was to come in from the northeast on foot and assist in the search of the objective buildings. There were only two – one mud hut and one shrine. While in Iraq we could not enter mosques unless we were escorted by an Iraqi. The Rules of Engagement stated that we could not fire at mosques unless we were being fired upon from them. They were no longer safe houses if the occupants inside were aggressive combatants. After the first bullet flies, they lose their non-combative status. This was the reason the ING were with us. They could enter mosques because they were Hajji and Muslim.

We positioned the Bradley to the north and dismounted to search the immediate area. To our right our other Bradley guys were providing security for the sniper team that was attempting to set up in the mud. Crowfoot, Pretty, and Crazy weren't having the best time. A call came over the net that another 113 had gone down. The second Bradley was dispatched to tow the downed vehicle on our return to CP 12. Like I said before, Murphy was everywhere.

The time passed like slow moving ooze. The rain fell like nothing could stop it. Mud was everywhere and it was as cold as I'd ever known misery to be. Before we dismounted to begin our assault, ODA and the ING had already secured the assault area. There was nothing there. The two buildings were empty and no caches were located. In his insistence that this was a legitimate operation and that Apache would leave its mark on history, Black Water Six continued to try to find something to bring back to Caldwell.

He was able to find a junk yard with a few more little mud huts but there were no caches there either. His Intel was either bad from the get-go or so much time had lapsed since he first got the info that the insurgents had time to come back and move everything they had stored up. Either way, nothing was there except a bunch of Joes who knew nothing would be there in the first place and that the operation was stupid for the amount of resources involved.

We moved in close to the shrine to rally with the other vehicles that were left behind as the CO went on his little adventure to the junk yard. When we were given the word that we were about to rally, I dismounted to inform the other vehicles around us that we were about to head out. The vehicles to my left had the ODA guys. When I went to their first vehicle, I found the driver asleep. He said that they had been going strong for a few days. He obviously didn't see or feel any threat at this location so he was catching up on sleep. It was kind of funny though; I felt a bit bad for having awakened him.

Our vehicles rallied and after deciding that the return route we had chosen prior to the mission was not going to be feasible, we headed back the way we came. This time our Bradley led the way. We moved along the tracks we had made coming down. At some point Priest lost sight of our tracks so Bong Water had to readjust our course. When he did this, he took us through the same wire that we had taken off the track earlier so we had to stop again to try removing the wire. While we were removing wire, the second Bradley was removing caked-on mud from the downed 113's tracks. Because it was being towed, a lot of mud had accumulated on the tracks and was making things very uncomfortable. We resumed our course to link back up with the Hummers. As Murphy would have it, we ran over more wire. It was ridiculous but strategically it would have been a brilliant idea if the insurgents had come up with it. Fortunately for us, it was just the farmers marking off their land.

We finally made it back to CP 22 and the hard ball. We were out of the mud fields. We resumed the lead and were able to move out at an exhilarating 10 mph. At one point in the tow we thought the other

Bradley was overheating due to the amount of weight it had been pulling, but once we got on the hard ball the tracks on the 113 were able to drop the majority of their caked-on mud. We took the speed up to 30 mph. Rather than spend more time switching tow vehicles, brass made the call to move on with the Bradley towing the 113. This was probably the best call of the day in that we were all ready to get back to base as quickly as possible.

We made it back to Caldwell with no further incidents but were cold, muddy, and pissed off upon our arrival – true infantry. We dismounted and unloaded our gear. The Bradleys went on to refuel. We all made it home in time for a hot meal at the chow hall.

****

From: <Malcolm>
To: <Mom>
Sent: Sunday, January 23, 2005

Hey Mom. It's been a few days since I've been able to get to the computers. We pulled duty at the mayor's office again and last night a big storm hit us so the normal place where I do Internet is down. But here I am. I got a care package from you today with the Star Wars sheet set in it. Thanks for getting those to me. They'll work real nice on my bed.

The Anti-Iraqi Forces have been doing a lot more now that the elections are getting near but we're handling everything professionally. It's a sad place. I'd hate to have grown up in a country like this. I'm glad and thankful that I'm an American and we have the liberties we do, even if some of them are screwed up.

January 24

Today was our day off. I slept until 0700. It was nice. I stayed in bed until about 1030 watching episodes of *CSI*. It was such a great show, something that I could look forward to when I had some down time. It made me feel like criminology was something that I would enjoy doing in life. I've always been intrigued by the process that criminalists use to investigate and determine truths about a crime. Forensics seemed to be what I gravitated towards but the whole idea of it intrigued me. I found that I really enjoyed the characters in *CSI*. Each of them brought something different to the table. I guess that's why it was such a popular TV drama – everyone who watched it could relate to a character.

I ran into Fillnutz at lunch. It was actually good to see him and he seemed to be doing all right although it looked as if his knee were still giving him some problems. He said it was but he didn't want to leave what he'd started here. Most guys would have hit the road by now. As bad as his knee already was, I didn't see how he had lasted as long as he had doing the things that he was doing.

I also ran into the Black Messiah at dinner. He was always in my prayers and thoughts. He said he was doing well at FOB Cobra although it was small and there wasn't much to do. He did say that he enjoyed going to their city's JCC. It was a welcome break from his life at Cobra. They stayed at the JCC for a week at time. He was the designated cook and all the guys looked forward to going there when he was slotted to go.

January 26

We didn't have anything scheduled so we did our normal maintenance day routine. A few of us cleaned the vehicles and the crew serves while the others sat around shooting the bull. That was the norm. The work was usually done in about 45 minutes but then we'd sit there for an hour or so more. We wasted two hours at the Bradleys just so we could say we wasted them together as a team. I guess it could have been worse. Worse was what happened to Deacon.

From the story I got, one of their Hummers hit an IED. It blew the front end off the vehicle. The gunner took the 240 to the chin and dropped inside the vehicle. DT said that he got out and ran to a ditch because he thought they might be caught in an ambush. From what others have told him, he went into shock. The driver was a little shaken up. The shotgun passenger, Gus, was hurt the worst. When the bomb went off, the metal in the vehicle bit into and crushed his legs, trapping him inside. Fortunately, the men on his team responded like professionals and were able to stop the bleeding and extract him from the vehicle with a little effort. After being extracted, he was attended to by the medics on scene and taken away on a dust off.

We heard later that he went into surgery at Anaconda but because of the extent of damage to one of his legs, he lost a foot. The pulse went out on the operating table. Feeling that the limb would no longer function properly anyway because of the mangled flesh, the doctors decided that this was the best course of action. Fortunately the gunner came out of it with only a cut on his chin and a lot of soreness. He was out walking around base later that day. (Update on Gus: he is fine and competed in a marathon at home only a few months after his injury. Thank God his attitude was bigger than his circumstance. He also came to Shelby to welcome home his platoon when they returned from Iraq.)

This morning's mission was to escort a vehicle carrying barriers to two schools to help establish security for the upcoming elections. It was not a glorious or romantic mission by any means, but it was one we had to do. We pulled up to the first school, dismounted, and set up a security perimeter. The dismounts watched as three guys unloaded and set up the barriers. Kids were everywhere. Every time we went out they loved to come out and see what we were doing. I'm sure kids were like that all around the world but in Iraq it was hit and miss for them to be around us. We never knew when we'd get hit, and if those kids were around us, they could end up collateral damage. Another thing that we worried about was that some kid or adult might walk through our area with a bomb strapped to him or her and blow the hell out of everyone around.

After downloading the barriers at the first school, we remounted and headed to an ING location to pick up more barriers. Since we had not been able to interact with the ING as much as we'd have liked, we took the opportunity to hang out then. They offered us cigarettes and we offered them a porno magazine when one of the guys asked if we had any "freaky-freaky" material. They were like bees to honey when we produced it for them. It was funny to watch their reactions. They swarmed the guy holding the magazine. We hung out with them for about 20 minutes, then loaded back up and headed to the second school.

The males seemed to be the most interested in us and said, "Hello, Mister." They always eyed us and would oftentimes wave. Kids asked for things like pens, pencils, money, water, food, or candy. Pretty always put on a good show for us when he interacted with any of them. He usually cussed them while he smiled at them, acting real friendly because he said, "They don't know what the hell I'm saying to them so I'm not really hurting their feelings." He also made friendly gestures to the girls and old ladies.

From what I could tell, once the girls here hit a certain age they begin wearing the dresses and shawls that cover most of their physical features. It was very strange coming from America where it seemed that every girl was trying to show as much of herself as she could get away with without being totally naked. That's what the Muslims were trying to avoid, I suspect. The less the temptation was, the less the opportunity for someone to fall to lust or even worse, some kind of criminal act. I wasn't certain what qualified a woman for the black phantom dress but I assumed it was worn when she became a wife. I later heard from one of the ING that it was worn to honor someone that she had known who had died.

We finished off-loading the barriers and returned to Caldwell in time to catch lunch. I then tried to take a shower, but once again, our water was out. It had been out for a few days. Thank God every one else stunk.

I went to chapel services. Going to chapel reminded me how much I needed to be thankful for what I'd been given, what I'd been kept out of, and where I was going. I'd slacked in my relationship with the Lord since this whole thing started. I prayed before and while on missions and at meals but the prayers often had no depth, just a quick blurb to ask for protection while we were out. I really needed to pray, read the Word, and more importantly, wait and listen for God to speak to me.

JANUARY 27

Over the past few days I've done some tattoos. I did two guys from the 386th out of Texas – Gonzo and Junior. They were good guys. I met them through another soldier named X. He was their sergeant. He brought them over to get the tattoos after I hooked him up with two. The two he got were touch-ups of previous tats. Gonzo got his name in old English across the top of his back. It turned out really nice. Junior got a U.S. Army tab with a KA-Bar knife going through it and Combat Engineer written small around it.

Last night I went to chapel with CFB. It was a good service. The guy who played banjo was there as was another with a mandolin. It was really cool but they didn't cut loose until the last song, "I'll Fly Away." Mr. Banjo seemed to be a bit more experienced than the rest of the band. I wished that they had let him lead throughout the service. He added a much better dimension to the sound.

Near the end of the service, in the middle of the sermon given by the Chaplain, Pretty came in and told me that we were being called out on emergency response to shots fired on one of the election sites and at the JCC. The JCC had just had a drive-by but the election site was under attack. It took us only 20 minutes to get fully loaded. We were on the road by 2130. Insurgents were brave when they thought it was only they and the IP; but when they heard us rolling up in our Hummers, Bradleys, or tanks, it was another situation all together. They might be bold enough to place a roadside bomb but they weren't stupid enough to take us head on. That was hard to deal with as a soldier. We wanted them to come out and fight us face to face, hand to hand if need be.

By the time we arrived at the site, the IP squad had fled and only the ING were there laughing about how the IP had handled everything. Major Ali was with the ING. I had heard a lot about him. Everything that I'd heard about him was true. He was a large man who immediately took control of situations. One thing I hadn't heard about him was his sense of humor. He was laughing the whole time we were with him. He and his guys were doing their jobs but they were having a good time at the IP's expense.

The Bradleys parked in the graveyard next to the school. After dismounting we began combing the area for insurgents. Our fire team crept around the dead in the full moonlight. It was a nice night to hunt Hajji. We entered a few houses and asked the people if they had seen or heard anything. As with most times when we soft knocked, the people feared for their lives and did not give us too much information. We continued patrolling around the city. Finding nothing, we went to the election site in the school.

It was here that we got the humorous side of Major Ali and the ING. They were laughing as they showed us where the IP had come under fire. They had fired back at the insurgents but they did not get any rounds over the perimeter wall. They hadn't even aimed high enough to clear the walls that surrounded them. After having a good laugh at this, we decided that it was an isolated incident and headed back to Caldwell.

No sooner had we left than the bad guys came back and did some more shooting. We heard about it on our way home but still managed to get back to base without incident. I decided to check my email and make some calls. Black Jack was the only one I was able to talk to on the phone. I told him what we'd been up to that night and then headed back to the CHU for some sleep. I got my pants off. That's when Pretty came by and said we had been called up to go out again and stay at the election site all night.

We reassembled and headed back out. We weren't too happy about the turn of events because most of us were ready to crash in our own beds. Now we were going to have to sleep either out in the open, in the Bradleys or Hummers, or not sleep at all and pull watch. We arrived at the election site and the ING were there waiting for us with big smiles. This time we found evidence of the assault. The insurgents had fired at least one RPG. We found the explosive but it hadn't detonated. Being good soldiers, we got pictures of the evidence for the Intel report.

While we were walking around the area, one of the ING was told to come to me because he had a gash on his hand. It was a puncture wound in the area of skin between his thumb and pointer finger, deep and ugly but not much blood. I cleaned it out as best I could. Then I wrapped it for him, all the while trying to communicate to him that he needed to get stitches as soon as possible. As I was giving him some medicine for the pain a bunch of his comrades came up and started telling me they had fevers. In reality, they just wanted whatever it was that I had given the first guy. It was funny. All the Iraqis wanted something from the Americans whether it was a piece of candy, money, or medicine. I finished wrapping the guy's hand and gave out medicine to the other guys. I then had to find a couple of Ace wraps for two of the ING simply because they thought the wraps were cool and wanted some for their own enjoyment.

While we were at the election site, we set up security on the Route C1: one Bradley to the north, one Bradley to the south, and Big Red's Hummer and PLS in the middle. Our gunners scanned all night. Some of us stayed out to help provide security while the rest of us tried to rest in the Bradleys. It was a cold night so rest was limited. We set up at 0345 and our plan was to wake at dawn, unload the concrete barriers we had brought with us on the PLS, then hightail it back to Caldwell for some much needed rest and hot chow. Murphy showed up.

We unloaded the barriers without incident and were told by the ING that some bad guys were in the town so we began a patrol. We mounted to go to the specific area. When we were turning around at the cemetery, our driver, Mickey D, took the turn too sharply in the soft dirt and threw the track from the Bradley. The track had come apart at a joint and that meant that we would have to replace the old one with a new connector pin. In the best conditions it was about a 30 to 40 minute job but when the unit was in the middle of a cemetery dirt road near a known hostile town and everyone was tired and pulling security, it took a few hours. Murphy was a pain in the ass. We dismounted to assess the damage. Big Red made the call to have the crew try to fix the problem while the dismounts went into town to see what we could turn up.

The town was just waking up and as usual, when the people saw that the Americans were coming they swamped us with, "Hello, Mister" and "Thank you." I wasn't really sure what they were thanking us for. They were so much nicer to us in daylight hours. We entered ten houses along our route, looking for suspicious materials and people. This part of town had some really cute little girls. There were also some older ladies who had tattoo marks on their faces. I didn't know what that meant in their culture. It was just one more thing that I wanted to understand about Iraq.

The patrol turned up nothing that posed a threat to us. We did find some pigeons, a dog foot hanging from a door, and some kind of deer head hanging from another door. Other than that, it was uneventful. We headed back to the Bradleys and heard gunfire in the far distance to the south. We took it as celebratory fire. We were not about to go chase it down even if the Bradley had been ready by then.

The Bradleys had parked on the west side of the graveyard. This was a pretty good location for us to gather. One of the mausoleums was open. While some of us worked on the Bradley, the rest took up positions in and around the structure. Some ate MREs. Inside the mausoleum was a coffin. Dark Wing and I each got a picture made while lying in the coffin.

January 29

Today was Election Day. We were on base, slotted with maintenance though we knew that we would be called out if anything bad went down. As of 1100 nothing had happened. Dark Wing informed us that some of us would be doing a mission to Ba'Qubah to transport vote cards. I hoped to go on that one. It would have definitely broken the monotony of our recent patrols.

Last night we did a patrol south. Actually, we just set up a road block. Our SP was 2000 at the Bradleys. We did our regular routine of roll call and then received the Op Order. The only person missing was the medic usually supplied to us by 1st Squadron. There must have been some confusion with them because OB said another medic was supposed to be doing this patrol.

We loaded up and headed west on Route T1 and then hung a left on Route C1. We reached the CP south of the election site where we had been two nights previously. It was a peaceful night. After dismounting we checked the area to be sure there were no IEDs or mines. During the day this wasn't as tense a job but at night it was hard to see what was on the ground. Fortunately the light from the night sky was good – just after the full moon. Twenty minutes after being in position, all the lights from the nearby cities went black.

We had nothing happening on the road so a few of us walked around in the dark with only the moon lighting our way. A person could have used Night Vision Goggles (NVGs) but the moon was so bright there was no need. I asked Pretty what he'd be doing if he was at home on a Friday night like this. He said, "I'd be in my truck with my gun, my friends … possibly my girlfriend … and we'd be hunting deer, listening to music and smokin' some bud. Yeah, that's what I'd be doing." I told him I'd be at Sunset Rock rappelling with my buds.

The nights we used to head to Sunset were just like this. The moon would be full and we'd load up and head to Sparta. Cigars and SOBE Adrenaline Rush were the norm. We'd get up on Sunset Rock and rappel for hours on end, just enjoying the thrill of going down the side of a rock Aussie style with only the moonlight to guide our way. After a few descents we'd hang out on top and maybe start a fire, smoke the stogies, and talk. It was beautiful. Sometimes we'd even con girls to get out with us. That was always a good time though most of them would never go down the rope.

It seemed like every time we were up there at night, we'd be visited by the local five-oh. The sheriff always patrolled the area for obvious reasons; people were out there at night doing whatever it was that people did. They usually weren't sober. We always were because of what we were doing. That was a good thing for us because the same deputy busted us for being up there at night. He was really cool about our being there but he had to do his job. When the 278th got to Kuwait, that deputy got assigned to the platoon I was in. Cruise Control turned out to be one hell of a good man. We were glad to have him with us.

Since nothing was going on at our CP we decided to take our fire team to a nearby house and search it. The family was really good about allowing us to enter and search. We did it quickly so we wouldn't bother the sleeping kids. After clearing the inside we headed to the back of the house. Pretty and I were poking around when a dog jumped out and started barking at us. It scared the shit out of both of us. We drew down on the dog and were thankful it hadn't been some insurgent because somebody would have died that night.

One funny thing that happened was when Boo Boo took a spill on our way back to the Bradleys. That was his third spill in as many missions. I didn't know if he was clumsy, a dumbass, or just careless. I'm sure it was more of his just being a dumbass. He had a tendency to get over-excited about things. He focused on everything else besides what he should have.

We stayed at the checkpoint until 2315 and then headed back to Caldwell to try and make midrats. It was funny to me how we so often planned our events around the chow schedule. I guess it worked though. It gave the patrol something to look forward to even though we'd bitch about the food once we got into chow.

January 30

Today has been a really laid back day. I got up at 0700 but didn't get out of bed until 1000, after a few episodes of *CSI.* At 1000 I headed over to Chapel. There were a lot of soldiers and KBR workers in attendance but the choir numbers were down; so were the numbers in the band. A quartet performed "Swing Low Sweet Chariot." It was interesting because they were all black and all I could think of was slaves out in the cotton fields or guys on a chain gang. I guess everything I've come to know about that song stems from some kind of black man's hard life and labor.

Last night we had a prayer service for the elections. It went really well. There were about 35 people present and we lifted our hearts and prayers to the Lord, asking Him to intervene on everyone's behalf in these elections. We were in the service for about an hour and I noticed that most of the prayers came back to the soldiers' needs. In this case, I didn't know if my thoughts had been tainted but it seemed that most often when we prayed or talked about God we always brought the prayers back to how He could benefit us, especially when we were praying for something as big as an election. I knew He was there for us and we could call on Him any time; but we were there for Him, for His purposes, His glory not our own. Many of the messages that I'd heard in the Church stroked the congregation's ego. The Church had to be big no matter the depth of their relationship with their King.

I went to chow with Logger Head. I didn't talk to him at all because one of the chaplains sat down beside me and we started talking about Chaplaincy in the Guard. I told him I'd been to the Assemblies of God Theological Seminary and that I had been ordained through another affiliation. He proceeded to tell me some of the steps involved in becoming a chaplain. I was praying about this last night and this morning. God was, is, and will always be faithful. I had no doubt about that. He brought the chaplain to me to answer some of my questions. Now the ball was in my court. I'd have to make a decision at some point and see how far I could run with the ball. I still had a heart for the elite units in the military. I still contemplated Special Forces. I wondered if there was any way to do both at my age.

After Chapel last night I came back to the CHU and watched *Platoon*. I haven't watched it since the first time I saw it in Cookeville when it premiered. There were a lot of similarities in the movie to what was going on here. I could relate to the characters Barnes and Elias. Both were seasoned soldiers but Barnes was the violent nemesis to Elias' crusader type personality. Somewhere in the mix were the everyday grunts. I'd seen both sides in Iraq and I suspected that was why I had been so upset while being here.

I told Boo Boo when we first got into Iraq that because of this war and our living situations, we would see the best in people and the worst. I needed to heed my own words. I've seen it in myself and I've acted both ways. My frustrations with some of the guys in our squad stemmed from my not recognizing that in myself until that point. There were going to be issues that I had no control over, that I wasn't going to like, and they were still going to happen. I just needed to learn to deal with those situations in the best way possible, by recognizing my attitude and dealing with that first and then dealing with the situation.

****

From: <Malcolm>
To: <Larry>
Sent: Saturday, January 29, 2005

Larry,

You're absolutely right about most people being good hearted given the chance, especially the children. I've enjoyed patrolling in areas where we're not seen much because the people take a genuine interest in us. Granted, we're always suspicious of everyone in this environment, but it's nice to be able to let our guard down sometimes and have a few jovial moments with the Iraqis. Most of them are really nice folks.

As for the beer derrick, I think that's a good idea. I'll try to come up with a design for that. I've got a friend in Cookeville who owns a rock/sand quarry so we could test it out there first, then just stay up there so I could feel like I'm back in the desert, right!

Yep, Mike's going to be the grand poo-ba again. How many times over does this make you a cousin? Will you dote on them the same way you used to dote on us when we were all younger? Ha ha.

Leap Year, My Ass!

February 1

This morning's operation was an ambush on Route C1 circa CP 10. Our wake-up was at 0100 but most of us didn't even sleep. We stayed up and watched movies, knowing that if we had lain down, morning would come too early. Roll call and Op Order times were slotted for 0200 at the Bradleys. At 0245 we mounted one Bradley and two Hummers and then headed out.

We got to the ambush site and set up our positions at five meter intervals. Our fire team, Pretty, Crowfoot, and I, was on the left with Boo Boo staying by Big Red so that he could have communication with both fire teams and the Tactical Operations Center (TOC). The other fire team was spread out accordingly by their team leader down the canal. We were all situated on the berm that was the west side of the aqueduct system that ran north/south of Balad Ruz. We planned to hold this position until sunrise.

It wasn't too bad other than the heavy winds and chilly air. Those weren't so bad because it wasn't really cold out. For the better part of the evening we lay in our positions scanning the horizon with our NVGs - waiting for some poor, unsuspecting Hajji to come by. The location of our ambush was a few hundred meters south of Summer School. It was one of the election sites in our AO. It was the site that we responded to when the IP came under attack, where we found the unexploded RPG round, and I got to treat one of the INGs for the puncture wound.

While we waited in ambush there wasn't much going on. We were basically in a position until we were given the signal to unleash hell or just do a show of force; rounds might not be fired at all. We were utilizing noise and light discipline. Unless it was an emergency, no light or noise was to be used. Our guys did a good job with it. The only light to be seen from us was when we turned on our NVGs. Noise was not so much of a problem because of the amount of wind around us and the sounds of water from the canal slapping against the concrete walls.

Our fire team, along with Crazy, was the response team should the opportunity present itself that a car came through our ambush. If they had their lights off they were probably going to be shot. We didn't have any activity for almost two and a half hours. Finally some cars came through. The guys were farmers and were getting an early start to the market with their onions.

When we did hasty TCPs (Traffic Check Points) one of us would aim our weapon at the vehicle from a distance. Another soldier would act as personnel overwatch. Another guy would frisk the people. The fourth guy searched the vehicle. We did this on three vehicles. The first and third went smoothly except that we had issues with the drivers leaving their lights on as we approached. I almost shot out one of the lights on the third vehicle but I kept thinking he couldn't understand me anyway.

On the second car we didn't know the guy had bad brakes, so when we signaled him to stop and kill his lights, he rolled a bit before coming to a complete stop. It was times like these that the pucker factor in Crowfoot and Pretty got high, especially in Crowfoot. They both got the rush but Crowfoot in particular got over excited. When the situation was over, they talked about what they "would have done" for the next two hours. Because we didn't know the vehicle's problems we had to assume the worst. Fortunately, as the vehicle came to a stop, the passenger was already on his way out of the door with his hands up. Crowfoot "would have" shot him had he not. Crazy and Crowfoot took the passenger's side of the truck while Pretty and I took the driver's side.

Pretty pulled the driver out of the vehicle and threw him on the ground. The guy fell to the ground and looked pretty helpless. Pretty gave him a little kick to let him know we meant business because he had failed to do what we told him. After a few minutes of seeing fear in his eyes, Pretty helped the guy to his feet and dusted him off. The rest of the search turned up nothing but onions in the bed of the truck.

That was our ambush. It made me laugh when I thought back to all the ambushes that I'd read about in Vietnam and learned about throughout my military training. In reality, all that we did was set up a disguised TCP. We moved the vehicles off the road and lay in the open on a berm. We started calling the TCPs Roach Motels. After the third car "visited" our Roach Motel and the sun started coming up, we moved out and headed to the JCC to make sure the guys there didn't need anything.

One of the things I realized about myself on that mission was that when I was in similar situations, where I was supposed to be alert and awake during extended periods of sheer boredom, I thought a lot. I thought a lot about girls – girls, girls, girls. I thought a lot about the girls I'd let slip by me. I thought a lot about the girls I'd never had the guts to pursue and I thought a lot about those fortunate girls I had been with (although I'm sure it was more fortunate for me).

I would consciously start out praying to God, hoping to have a good quiet time with Him while I was on watch or whatever, but I would always end up fantasizing about girls. I came to realize that my mind would not afford me the luxury to spend time with God while I was out there pulling a guard. I had to stay awake and for some reason, the thing that kept me most alert was thinking about those girls – explicitly fantasizing about those girls.

February 2

Bore and Johnny Vega showed up at Caldwell today. We saw them at lunch and invited them up to our CHU for a little family reunion. Bore and Vega were in I Troop with us before we hit Kuwait. They were both in 3rd Platoon. Both were on their way to R & R back in the States. It was good seeing them because the majority of I Troop had recently been at the site of a suicide bombing. They were pulling JCC duty in the town that their company patrolled when it all went down.

A suicide bomber had come to the JCC to kill the mayor of the town of Khanaquin during the elections. Fortunately for our guys, the ING who were frisking people that day at the entrance of the JCC did their job well. They paid the ultimate price for their competency but they did their job. They didn't allow the bomber in the area because he looked suspicious. He left the area for a few minutes and came back with a little girl. Then he blew himself up where he was, killing himself, the little girl, three INGs, and two other civilians. No American soldiers were in the kill zone.

Iron Troop had an incoming patrol, led by the CO, which had to park in a spot not normally used as their parking area at the JCC. New concrete barriers had been put in place to provide additional security to the civic building. The barriers, in a roundabout way, assisted in saving a lot of lives - Iraqis and Americans in the building and Americans who were coming to the building. The soldiers who were in the building were standing in locations far enough away from the blast that all they felt was the concussion. It did scare the shit out of them though.

Ragin' Cajun was the medic on duty that day and was probably the closest to the blast. From what the other guys had told me, he was pretty shaken. In a split second everyone there went from normalcy to extreme trauma. The dead and dying were strewn about and all that could be done was to try to figure out which bodies were still breathing. One could tell that some were alive as their screams of agony and disbelief filled the air. Others were dismembered remnants of what the human body was supposed to look like.

Pieces of the bomber's and victims' bodies had gone everywhere. A piece of someone's skull fell near Vega. Brain matter was on the walls of the JCC and on the sidewalk. A cat thought the meat would make a tasty lunch. Blood and singed flesh hung from nearby trees and power lines. The face of the bomber, having been blown off and somehow having stayed in one piece, was found on the ground a few meters away. In a daze, Cajun attended to the bodies. There was very little he could do as the majority of those in the immediate area had been killed.

Our standard operating procedures for a situation like that was to clear the area immediately, set up security in case of any secondary explosions, and then wait for EOD to respond. With casualties involved, things took on a different direction. Triage and mass casualty procedures had to go into effect. Our soldiers were professional and responded well. Fortunately there were no other bombs or bombers. I Troop secured the area and rallied to get things cleared up. Afterwards, all the guys who had been there received post-trauma psychological debriefings.

FEBRUARY 4

Last night the Double Deuce came under direct enemy fire. We were south of Balad Ruz, patrolling the same little town that we had been in numerous times before. We dismounted a few clicks north of town. I didn't mind. It was a nice night. We patrolled down Route C1 until we hit the outskirts of the town, where we turned right onto the main dirt road that ran through town. We patrolled the side streets and goat trails in and around the town with no incident except the occasional barking dog.

Pretty was walking point, followed by Crowfoot, Dark Wing and Boo Boo (with the radio), me, Little D, Lumberjack, Crazy, and finally, Spider. We had just scaled a wall and were in a little ditch system, which ran along the southwest edge of town, when Pretty called us to a halt. I whispered to Little D that this area would be the perfect place to get into a fire fight because we had such good cover. We could duck and run and the only real advantage the opposition would have was that they knew the area better than we did. A few minutes later Pretty and Crowfoot came back to our position and said that they had seen a couple of guys up ahead, roaming around the area with guns in their hands.

We slipped quietly back into formation and were rounding a corner when shots rang out. Pretty, Crowfoot, and Dark Wing began returning fire. Boo Boo ran for cover when he heard the shots. I squatted for a second to get my bearings of where the shots were coming from and then headed up to the front. From what I was told later, Little D,

Lumberjack, and Spider hit the deck quicker than diarrhea on a bad day. Crazy had to hurdle two of them to get up to our position. He got to our position and began firing illumination grenades.

As I was running up I heard Dark Wing shout, "Get up here!" to the rest of the squad. By the time they did, the gunfire had ceased. I circled in front of Pretty, Crowfoot, and Dark Wing to take point. Thinking back, it probably wasn't the smartest thing to do because I crossed directly in front of their field of fire. They seemed to be in some kind of bizarre conversation about who and what was out there so I thought it best to be in front, in a secure area, out of everyone's field of fire when I got set.

I shouted behind me, "Where did the shots come from?"

They yelled back, "The shots came from the berm in front of us!" I started leading the squad in that direction. When everyone else had regrouped we sprinted to the top of the berm to look for the shooters. The squad crested the berm with Pretty, Crowfoot, and me in the lead. We had no sooner gotten on the berm than Crowfoot fell into a hole and twisted his knee. After getting him out, we scanned the area and found it full of craters, ravines, and ankle-breaking holes. It definitely was not friendly terrain. The teams fanned out and searched the immediate area for any evidence of the shooters. We were able to find the area from which the insurgents had fired; spent brass was on the ground. Pretty and Crowfoot said that they had visual on three individuals prior to and at the time of the shooting. Pretty had spotted them through his NVGs.

Our search did not turn up any bodies. We counted numbers and then headed back to the vehicles. Once there, Gunro said that through his infrared sights he had seen a body enter a house. We dismounted again and headed back to the area we had just left. We found nothing out of the ordinary and assumed it to be someone just curious about what was going on outside his home.

We returned to the vehicles once again, mounted, and then headed back to Caldwell. Fortunately, no one had gotten seriously injured. Crowfoot complained about his twisted knee but he knew it could have been a lot worse. I treated him once we had gotten back to our CHUs.

Something about it was unsettling to me. I guess I'd watched too much *CSI*. Since I did not get the opportunity to fire my weapon or see any hostiles and because I'd had time to think about it, I wondered if there were actually any gunmen on the berm in the first place. These were the kind of thoughts that went through my head during down time. Pretty, Crowfoot, and Dark Wing had been in tight quarters a lot recently and they could have set the whole thing up so it would seem like we had finally gotten into some action. I wouldn't put it past them but in the same sense, why would they? I had too much time on my hands.

I thought Pretty had picked up some brass from the Summer School shooting at my suggestion. I had told him that it would be good for just that purpose – to have evidence of a fire fight should we ever need it to prove we were in one. We only found those shell casings because they were in the school and the ING had showed us where they were. Every other time we had looked for brass we had not found any. This time we did find brass. It would have been easy to set up, stupid, but easy.

I hoped I was wrong but one never really knows unless one sees the muzzle flashes, hears a zip or ding, or feels a round ripping through his flesh. Regardless of what really happened, it was good to know none of us had gotten hurt.

Yesterday morning I volunteered to go out on patrol with Black Water Six. We were told to meet at the motor pool for our Op Order. Three Hummers and a 113 were to head out together with BW6, First Sergeant, Legal, and PsyOps (psychological operations – the ninja mind trick guys). Our mission was to take BW6 to the JCC for a routine town meeting and then, time permitting, drop some flyers with the PsyOps guys. Legal was going to be dealing with claim issues that the people of Balad Ruz had while BW6 was in his meeting. The rest of the element just sat out in the parking lot or hung around in the JCC, waiting until the meeting was finished.

I loaded into the PsyOps vehicle along with an Iraqi interpreter named Jacob. Jacob was from Baghdad but hadn't been there in three years due to his job and the threat it would present to his family should an insurgent find out who he was and what he was doing. Despite all that, he really seemed to enjoy his work. He was good at his job. He was a neat guy – very friendly.

Our march to the JCC was interrupted when Black Water 6 wanted us to check out a couple of vehicles parked along Route T1. We checked the vehicles and began to move away from the area when someone spotted some suspicious holes dug along the road. Freshly moved dirt was always a big "Caution" when we were on a road. We never knew if someone had laid a bomb underneath. BW6 and a few others dismounted and began walking along the road to inspect the holes. We followed slowly in the vehicles. After checking most of the holes, the team mounted and continued its mission to the JCC.

While I was in the parking lot, some of the ODA guys arrived. They pulled up in a nondescript Hajji van and a similarly, ordinary car. Both vehicles had their windows tinted so that from the side, no one could tell Americans were inside. The only tactical gear that the guys were wearing outside their street clothes were bullet proof vests. They all carried machine guns and wore no helmets or uniforms that might say, "Hey! I'm an American. Shoot me!" They looked focused but very relaxed, and they seemed intent on getting their business done at the JCC quickly before moving back out. They weren't about fooling around.

While there, four of us went to the ING station to do some training with their guys on medical situations. We were invited into the office of Major Ali. Six other ING were in the office including an interpreter. Major Ali had recently arrived and had yet to change into his uniform, so while we were talking to the rest of the men, he changed. Once dressed, he left the room and came back with some "snivel gear" that was normally issued to his men. He gave each of us American soldiers a pair of long underwear type shirt and leggings.

Three vehicles were sent back to Caldwell because Jacob had seen a man on the way out who had a beard the day before and was clean shaven on this day. That wasn't normally a big deal, but in Iraq it was an issue because more often than not, a man who recently shaved his beard was about to do a suicide run. This guy was on Caldwell at the moment. The Regiment finally heeded Jacob's warning so we loaded back up and headed to base. Upon our arrival both Jacob and the XO went into the squadron TOC to brief the Squadron Commanding Officer (SCO) about what Jacob had seen.

After doing business at the TOC, we headed back to the JCC. We waited a short time for the meeting to end. When it finally did, we mounted and headed back to Caldwell. We had just finished clearing our weapons at the barrels when Black Water 6 announced that we were going to investigate an explosion north of Balad Ruz. We turned around and headed back down Route T1 and then north on Route C1.

The convoy came to a stop in a field where we could get a clearer view of the smoke in the distance. The smoke was so far out that we realized it was not in our AO so we mounted back up and patrolled some of the adjoining fields before heading back to Caldwell. We returned to base around 1645. It was a long morning and I was slotted to go out again at 2000.

****

From: *<Malcolm>*
To: *<Distribution List>*
Sent: *Saturday, February 05, 2005*

Hey folks. Just a little update on what's happening here in good, old Iraq. As you know, the elections are over and thankfully, they went better than we expected. I don't know what the media has put out but we were really happy that the elections went off without too many crazy incidents of suicide bombers or vehicle-borne IEDs. The people seem to appreciate the fact that they now have a chance to have a say in government. Hopefully they won't abuse it like we sometimes have a tendency to do.

We've been "in country" for almost two months now and almost three since we hit Kuwait. It's been interesting to say the least. Our days are filled with patrols, eating, working out, and Internet access. It's almost like being at home but not. I've been doing a few tattoos as well and that's been one of the highlights of being here because I'm meeting a lot of soldiers I probably never would have come across otherwise – most of whom are from a detachment out of Texas.

Our squad, which we call the Double Deuce, was detached from our home company to be with Apache Battery of the 1st Squadron, 278th Regimental Combat Team. It's been a good thing. We are a mechanized infantry unit which means we ride out on Bradleys [and] then dismount [to] do foot patrols throughout the area. Sometimes it gets old but then we have nights like two nights ago when nothing happens and we get to enjoy the stars and palm groves. It's really beautiful here on nights like that.

Right now our squad is in the process of getting custom rugby jerseys made to commemorate our time here, so beware when we get back because we're not really a rugby team. I'll send a picture of the design once I get it drawn up. It's pretty cool.

A lot of you have been kind enough to write me almost daily with emails and more than I had expected have sent snail mail and packages. Thank you all for that. It's such a good feeling to know that while I'm here you are here with me and support me however you feel about this war.

Actually, we are right being here and we're doing well for the country. It's a long process that the Iraqi people have to go through, but I'm confident that there are more good people here than bad and they will overcome and prosper. Now I realize why my cousin Mike was so proud of what he was doing here at the beginning of the war with the 4th Infantry Division. My hat's off to you, Cuz.

I don't know how long we'll be here other than when the higher ups say we can leave or when Iraq feels like we've been here long enough and they are ready to run their own country. Hopefully it will be sooner than later but I plan for the long haul, so keep the letters, emails, and the care packages coming when you feel inspired. I love hearing from each of you. Makes me feel like I'm still part of your world. That's about it for now. Hope all's well in your world cause it's cool in mine.

February 6

Yesterday we spent an hour doing maintenance on the vehicles and weapons systems. The rest of the day I did tattoos. I did a multi-colored scorpion on a KBR guy named Mike who worked in Bradley maintenance. Bong Water had "OIF III" inked on his right upper arm. I put Chinese letters and a tribal on Gentle Ben. It wasn't a bad day money-wise but I should have charged a little more than I did for the work.

February 7

Today I got firsthand knowledge of what an IED could do to the human body. Our patrol started sometime around 0400. We headed south on Route C1 to the area where we had come under fire a few nights before. Our orders were to set up a Roach Motel in the area, questioning anyone breaking curfew. While moving down the road our lead vehicle noticed a recent blast hole in the route. The gunner also said that he saw what he thought were a couple of bodies on the side of the road. We did not stop. Big Red decided we would check the area after we were done with our Roach Motel.

Roach Motels were simple operations. The local farmers were out early on their way to the market. And what was Joe going to say to any of those guys just going out to do their jobs? The market was their livelihood. At daybreak, we loaded up and headed back north. A call from the lead vehicle said that there were two bodies at the blast site reported earlier. We immediately set up a perimeter and checked the area. I saw men but the reality was this: one and a half human bodies were on the side of the road. We suspected that they had been in the process of setting up an IED when it had accidentally detonated and killed them. I thanked God it happened to them and not us.

Both of the men had been dead for a few hours and rigor had already set in on their bodies. The closest man to the road had apparently been the man farthest from the blast, maybe standing behind the guy setting the bomb. Not only did the concussion slam his body but he was rippled with shrapnel holes. He might have lived for a little while after the blast because of the way he was clenching his hands, but there was no doubt that he hurt like hell on his way out of this world.

Doom on you Ali Babba for trying to set an IED

His face and legs were torn apart. One knee was completely blown out and bone marrow dripped from the wound. The man's face was like that out of a Rob Zombie horror flick. It was burned from the blast, but the most haunting part of his face was his eyes. All we could see were his sockets, caked with dirt and filled with blood. There were no eyeballs. Panic and fear had signed their names on his soul before he died. His mouth was in the death scream position – pain was his epitaph.

The other man appeared to have had a much quicker but uglier death. He was definitely the one closest to the blast. Either the blast itself or some kind of shrapnel had taken his head off completely. There was no evidence left of the type of ordinance the guys had used, so we couldn't tell what exactly had, in fact, decapitated him. All we knew was that the only thing left above his neck was the top of his scalp connected to his upper torso by a thin piece of skin. Both of his arms had been severed and were contorted in some vaudeville-type position as was one of his legs at the knee. He was a lump of mangled flesh. When we stood where his head was supposed to be we could look down at his body and see clearly into the cavity at his heart and lungs. Half of his nose and mustache that were ripped off in the blast were a few feet away as was some of his brain. The rest of his him was just a lump.

Both men had lived violently and had died even more violently, although I believe the decapitated one never knew what hit him. They were the epitome of the devil's rejects. "So much for seeing heaven," I thought as I looked over them. After securing the area and calling the ING, IP, and EOD we maintained distance to await their arrival. Traffic started coming our way from the north and south. When IP and ING did arrive, they each checked the bodies. EOD arrived last and did their business in establishing that there were no other bombs placed in the area.

When EOD had finished their bit, a few of us returned to the site to search the bodies for any evidence, identification, and/or weapons that had not been destroyed. We found money, wires, ink, and a pistol with a full magazine. There appeared to have been at least one other person involved because we found an AK magazine on the first man and a blood pool on a drive way 20 feet down the road but no AK-47. We assumed the third guy must have picked it up after the explosion and headed out as quickly as possible – wounded but alive to fight another day.

An ING searched the bodies last, and when he was through, he gathered up all the evidence that we had already photographed and logged and took it with him. I assisted the local ambulance driver in loading the bodies. I had no clue as to what they would actually do with the bodies. That was their deal so we were good just getting them out of the area. Looking into the man's eye sockets while I was loading him was a bit eerie. Mortality and all the horror movies I had ever seen flashed back into my mind. I wondered when he was going to jump back up and try to kill us all; but unlike horror movies, this was the real deal and this guy was deader than driftwood.

The thing I noticed most about the headless guy when we loaded him was the stench that surrounded him. Death has a bizarre smell and even more so when mixed with singed flesh. It was an awful smell that I probably won't ever forget. It smelled like shit – nastier than we had dealt with in town in the ditches. It was foul and I assumed it was from defecation or from his bowels that were hit. Whichever place it was from, it was bad. I hoped I'd never smell it again. Once a person has something like that registered in his senses, any other time it hits his senses, it's going to be a reminder of that first incident. I've already smelled it a couple of times since. Maybe it wasn't from the guy's body at all; maybe it was just Iraq.

FEBRUARY 8

Today our patrol started at 0400. We were slotted for a raid with Major Ali and his guys. When we arrived at the motor pool we were told that White and Green Platoons had both been out all night chasing bad guys. Green had gotten two of its tanks stuck in a mud field. "Some studs," we thought, "if in their infinite tactical wisdom they had gotten their tanks stuck in the mud." The goobs shouldn't have gone through the mud fields in the first place. I guess they didn't consider the weight of the things. Our mission changed slightly in that we were going to escort a recovery vehicle (for towing purposes) and a refueler (obviously to refuel) to the tanks' location, drop them off, and then proceed to FOB Knot for our 0500 rendezvous with Major Ali and the ING.

We dropped off the two vehicles and linked up with our counterparts near CP4. Next, we proceeded south on Route C1. Our plan was to raid the suspected house of one of the insurgents who had been killed the previous day in the IED placement. We had found his identification when we checked his body.

Historically, communication was an issue in every military operation that has ever gone down and it was compounded exponentially when two forces, working together, did not speak the same language. Some of our ING counterparts could speak broken English but most of the Joes could not speak a lick of Arabic. This proved to be the most difficult factor in our venture to the raid location. After some negotiating on which house it was that we needed to enter, we drove to that location and began the raid. Nothing turned up. At daybreak we headed south again to recon more of the area.

We were about thirty minutes into our drive south when a vehicle started firing at our column. Our vehicle was last so most of what we saw were tracers flying to and from our left side. Big Red came over the radio and said we were being engaged and were going to pursue the vehicle. The Hummers and trucks picked up speed to close on the vehicle.

The ING led the way as they were the ones who were closest to the suspect vehicle. It was like a *Dukes of Hazard* episode with the kind of driving we were doing. We raced down muddy rice paddy roads, beside and over aqueducts, sometimes coming close to rolling into the water, sometimes coming close to landing in ditches, and all the while, dodging the holes and ravines that run throughout the country's side roads. Fortunately God was watching over us and I was a pretty good driver.

The ING guys finally apprehended one of the suspects. For some reason, the shooter had jumped from the pursued vehicle and it had gotten away. The ING guys were on the insurgent like a swarm of angry hornets and interrogated him on the spot in the "way they know how." They got the information out of him pretty quickly after "tenderizing" him a bit. Though I couldn't see everything that was going on from my location in the convoy, I knew Major Ali was establishing his authority with the man. To finalize the idea in the man's head that the ING hated the insurgents and all they stood for, they fired off a few rounds near his head to scare him to death.

We detained the man, loaded him, and then headed to the area where the detainee said his friend was hiding. We searched one area and then another, finding nothing. During our second search someone found us. We were the last crew and were pulling rear security. Scooter, my gunner, spotted a truck heading to our position. My instincts said to turn our Hummer around so we could be ready to move out quickly. As

I started to move, we heard automatic gunfire. I got on the radio to tell the other vehicles that we were turning around to see what was going on. As I was turning I saw all our dismounts running toward the road on which we had come in and our coalition team was firing on the red truck that Scooter had spotted moments earlier.

One ING truck began pursuit. I stopped long enough for Crowfoot to jump in the vehicle and then we closed in behind them. Rain was coming down and we were on mud roads, so the driving became even more intense as we slid in the mud around curves and in the tracks of the two vehicles immediately in front. We hit hardball for only a few seconds, then immediately turned west and made another quick left to drive another mud road along the south side of an aqueduct.

We had eyes on the truck and the ING were shooting at the suspect. The whole time we were speeding along I was yelling at Scooter to fire the Mark 19 grenade launcher if he got a clear shot. He never fired, so I assumed he never had a clear shot or he just never was able to because of the rough ride he was getting. After a few minutes of high speed chase, the red truck got stuck in the mud and the driver jumped out, trying to make a run for it on foot. He took a bullet in the right pelvic region and went down.

Our vehicle arrived on the scene seconds after the runner was gunned down. Our immediate problem was finding a way to get to him in the field as it was so muddy and we really didn't want to walk any more than we had to. Scooter spotted a route that we could utilize to get over the berm that separated us. We'd be able to navigate to within 30 feet of the casualty. Georgia, another medic, caught up with me as I trotted over to the Iraqi to examine the extent of his injury. We found that the bullet had entered the pelvic area and it appeared to have been the perfect spot that cut the femoral artery. There was no exit wound. The guy was bleeding out like a cow when it gets its throat cut in the slaughter house.

Georgia began to apply direct pressure on the wound as I started an IV. We applied Curlex gauze first and then tried to apply an Israeli bandage to secure the Curlex in place. Because of the area where the wound was, the bandage did not work as well as hoped for and he continued to lose a lot of blood. With one IV started to replenish his fluid loss, I began working on a second in his other arm. After a few minutes I had one of our guys hold the IV so that I could assist Georgia. We traded places so that I could continue applying direct pressure and he could get some Quick Clot, a fast acting clotting agent that was supposed to stop hemorrhaging. It did not in our case. It worked like magic in the training videos we saw in school, but it did not work here.

Georgia and me working on AIF soldier

The insurgent's blood pressure and will to live began to drop quickly. We decided that our best bet was to continue direct pressure since that was the only thing that seemed to be working in stopping the blood from gushing out. All said, we gave the man five bags of fluid though two arms. We were finally able to put a tourniquet on his leg. It was placed as high as we could get it. In reality, it was directly over the wound to hold the pressure so we could get our hands free to work without constraint.

His pulse was rapid but weak and his respirations were labored. We saw his eyes getting glassy and he kept trying to cross his arms and go to Allah. He wanted to die. Shock had hit him and if we didn't do something else we were going to lose him. We put him on a litter and elevated his legs so that the blood he did have left would go back into his core. That did the trick. His color came back and he become more reactive to the things we were trying to do. Nobody had a wool blanket in his vehicle but I did have a space blanket in my aid bag so we wrapped it around him.

While we were administering aid, Crowfoot radioed the 9 Line, requesting a medical helicopter to come to our location and "dust off" the casualty. The information given in the report gave the pilots an idea of what was happening on the ground and the flight nurse enough head's up to handle whatever casualties were coming aboard. It also gave the hospital a head's up as to what kind of treatment needed to be administered when the casualties arrived – what to expect. Finally, it was an opportunity for a medic to be re-supplied in the case of multiple wounds or mass casualties.

Me making sure the AIF soldier doesn't bleed to death. He didn't.

A bird arrived 45 minutes after we got the go-ahead from Baghdad. Red smoke was popped to show the bird where to land and what the wind direction was. It landed and the flight nurse dismounted. We told him what we'd given the guy as far as aid; then he led us and two other litter bearers carrying the insurgent to the bird. We loaded him on the bird and the flight nurse boarded. Atlas flew out to provide security for the flight crew. They headed to Baghdad for the insurgent's advanced treatment.

We did a good job that day. We saved someone's life. It was a good feeling for me although the ING couldn't understand why we had saved the guy. For their part, he was an insurgent and he should have been left to bleed. What they should have done was finish him off right there, before we arrived. But they didn't. I'm glad we got there when we did. If we had not, that man would have died in that muddy field, maybe deservingly; but we were obligated to do our job and provide the best treatment that we could.

After ensuring that all the medical trash was picked up and we congratulated ourselves for the way we all worked as a team, we mounted up and headed to the hardball. Our mission was finished for that day. We made it back to base just as our security guy, who had flown out with the insurgent, touched down at Caldwell. The man had survived the flight and was stabilized until the medics in Baghdad took him off the bird. The bandage and tourniquet, after all the jostling around, had come loose. When it finally fell off he began to bleed again. I'm sure the Quick Clot was burning in the wound too. The last we heard of him, he was being rushed into an ER screaming.

I was in Caldwell long enough to grab a quick lunch and even quicker nap before heading back out on another patrol. We headed south again to set up an ambush and Roach Motel. We were in the area for a few hours when our Bradley commander reported that he had picked up some people in a field through his night sights. Our guys dismounted and went on a foot patrol in search of the suspicious characters.

They came up with nothing and returned to the vehicles. One of the good things about driving was that we didn't have to do the foot patrols that turned out to be a bunch nothing. This turned out to be one of those patrols. Gunro was sure he saw movement. We didn't argue with him. He had thermals. What he didn't have was a clue as to the distance on the targets. That's what became frustrating when he was out with us. Sure he could pick up possible targets but we wasted a lot of time searching on foot because he couldn't communicate where he had picked them up.

****

From: *<Winston>*
To: *<Malcolm>*
Sent: *Tuesday, February 8, 2005*

I am a little late, but I want you to know that I just mailed three care packages (total weight of about 100#) for you and your men. There is a lot of beef jerky, Slim Jims, chewing tobacco, and paperback books.

Kaye said that the look on the face of the lady at the check out line in Costco's was priceless ... The checkout lady couldn't figure out why this crazy woman (Kaye) was buying this huge supply of jerky and chewing tobacco! Kaye started to clue her in, but then decided to have a little fun. So, she said nothing and left the poor lady dumbfounded.

FEBRUARY 10

Dark Wing, Crowfoot, Nasty, and I provided security for the 1st Sergeant's patrol to the JCC for the weekly meeting of officials. While we were there I taught a first aid class. Twelve guys sat in on the class and were really good sports about listening to me through our translator. One of the guys allowed me to stick him with an IV so I could show the others how to do it.

I also got to see Major Ali. While I was in his office I was treated to a snack of bread and jam. An hour later the ING invited me to share lunch with them. The meal was rice, bread, and some concoction of beans. I finally got to see how their bread was made – pretty interesting.

The dough was mixed in a deep dish and then rolled out. Next, it was put on some kind of padded hand tool that was slapped against the side of the oven. The dough would stick to the side. The oven looked like a hornet's nest turned upside down. The opening was big enough to put one's head in but obviously that wasn't something to do when the fire was burning. It was fueled by some kind of propane or natural gas. With the dough stuck to the wall of the oven and the flame keeping the heat at a good baking level, it only took a few minutes for a pizza-sized bread snack to be ready to eat.

After lunch, we walked back over to the JCC. We arrived to find a crowd of people around the body of a dead man who'd been shot by insurgents. He was young and had apparently been executed for working one day of the elections. The AIF continued to be serious about not wanting anyone in power but themselves or their puppets. Major Ali was above the body of the man, and when he saw me he asked for my camera and something with which to cut the clothes. I gave him my camera and a pair of scissors. He cut the man's shirt and took pictures of all the bullet holes. There were two in his back, one in his chest, and one in his abdomen. His face was bloodied as if some other form of trauma had happened there; but it wasn't crushed or beaten so we assumed either he had gotten shot in the head, butt stroked, or pistol whipped. The saddest part of the situation was watching the father of the man try to find some justice in it all. He kept searching for answers that no one could give him.

We were ordered to mount up about 30 minutes later. Major Ali was going to lead us down south to try and stir up some answers to the man's death and the father's questions. We followed Major Ali and the ING south to search the area for AIF. They found nothing but did stop a few cars. We were the last vehicle so we did not do much in the way of taking an active role in searching. We were along for security and force protection.

We finished our search down south and headed north. When we got to Summer School we got a report about a blast to the NW but we did not investigate it. The last few times we'd had reports of explosions we had ended up chasing ghosts, so Black Water 6 decided we'd just let this one go.

February 11

It was the second of our maintenance days. We were on green cycle and were slotted for a day off but were told we had a 1300 meeting. The squad leaders had something planned but would not tell anybody what was going on. I hated meetings. They were so long and more often than not, not very productive.

I also checked my email. To my surprise I had one from Squiggy, who just happened to be one of the most gorgeous, God-fearing women I'd ever met in my life. I met her in Springfield, MO while I was at seminary. I also got one from the Hit Babe, who also happens to be a gorgeous, God-fearing woman. I had talked to the Hit Babe a few times since leaving Springfield but hadn't talked to Squiggy at all since leaving. It was a special treat to hear from them both on the same day.

I also did a couple of tattoos that afternoon. I did a tribal piece on Boo Boo and barbed wire on a guy named Gram. I was going to do one on Carol but my time for doing tats ran out. The other two tattoos took a few hours and by the time I got around to doing his, it was getting late in the evening and we had an early morning ahead of us.

We heard reports and saw the news coverage of a VBIED that had detonated in Balad Ruz while Red Platoon was out. We had learned from Red that three ING were killed and one was critical. Many civilians were killed as was the driver of the VBIED. A second VBIED, the one that the ING initially investigated, did not go off. It was parked across the street from where the ING were observing. As the ING were clearing the area, the suicide bomber drove into their location and detonated his bomb, instantly killing him and many innocent Iraqis. It was just another incident of these assholes coordinating attacks of senseless violence. There will always be death in war and I can respect a man for being willing to die for his beliefs. What I can't respect is when he thinks that killing individuals who are his countrymen - women and children who have nothing to do with the war and taking his own life in the process somehow makes him a martyr. There is nothing honorable about that course of action. It is terrorism – nothing more, nothing less.

We also received reports of a sniper who took shots at the JCC .50 cal position. Big Mac was on watch. Fortunately the sniper did not hit Big Mac or any of our other men standing watch down at the mayor's office.

I started reading *The Godfather*. It was really good. I had always planned on seeing the movies. I used to tell people all the time when asked if I'd seen the movies, "Yes. I've seen one." I had not watched the trilogy. Buzz sent me the book in one of the packages that was sent by "The Keebler Elves." The Elves were really good about taking care of us. They sent Levi Garrett for Dark Wing, Copenhagen for me and Marlboro Lights for some of the other guys, some books, and some new music.

February 12

Morning came early. We were up at 0420 for a 0500 rally at the motor pool. Our patrol was in the northern section of Balad Ruz. Coming in from the east, we knocked on a few doors, woke up some of the locals, and then moved on just north of the JCC. We did another foot patrol for a short time and entered a few more houses.

In one of the houses we found a few AK-47s. The guns were authorized for the amount of males in the family so I had a picture made of me posing with them. The family laughed a bit as they realized what I was doing.

For some reason, one that we determined was simple vagitis (that's Army lingo for being a pussy), Shrek didn't join us on JCC duty. He would have been the ranking NCO but in his absence I became the senior man. It was no big deal. Things went just like they always had. We tried making the watch rotation a little better so we could enjoy more time off. Not much went on while we were there. Eventually though Dark Wing made the call to send Crowfoot down to run the watch. It was a slap in my face. I was every bit as capable to run the watch as any other NCO.

We witnessed the IP stop a car and then whip the driver's ass when they pulled him out. It was over quickly but definitely livened up the day. Sometimes things like that just made us glad we were around to watch. The rest of the day was uneventful – boring. I was able to get a kabob and some bread from the cleaning boy.

****

From: <Malcolm>
To: <Becky>
Sent: Sunday, February 13, 2005

Hey Beck,

Things are good here. I'm on guard duty at the Mayor's office in the town we're charged with. It's not too bad as you can tell. They've put an internet line in here for us so we bring in our laptops and surf the net and check email.

I saw a bunch of death this past week. A few gunshot victims and two explosion victims. It's not bad because they're nobody I know but weird just the same. The explosion guys were the worst because they were so surreal. I got some pictures of them before we hauled them off, but I don't think you need to see them. They're pretty graphic.

Thanks for praying for me. I appreciate it. It's nice to know that our relationship has grown like that over the years.

FEBRUARY 14

Boo Boo paid me $65 to stand his watch from 2300 to 0200 this morning. I had the watch after that too. He did not want to go up on the roof because he thought it was too cold. I liked it. It wasn't as cold as I thought it would be and going up top gave me an opportunity to think about our situation in Iraq, God, friends and things we used to do together, and girls. Girls flooded my thinking. I had to laugh at myself sometimes because I'd been ignorant in most of my relationships with women – always turning women away from me because I was so scared of hurt, theirs and mine. I would rather not take a chance at getting hurt than take a chance at falling in love. I missed out on a lot of experiences I might have had otherwise.

We were relieved at 0830 by Red Platoon. I ended up playing Mark 19 gunner and Track Commander of a 113 on our way back to Caldwell. I'd never manned the weapon so when I checked the loading of the gun I found that I couldn't get the cover to close again properly. I rode back to base with an improperly working weapon. If anything had gone down I would have had to use my M16.

At the clearing barrels we dismounted and checked our weapons, making sure no rounds were in any chambers. I got out of the TC hatch and made my way to the front of the vehicle. As I was getting ready to get down from the deck of the 113 I had one of those "Oh my God, I'm about to fall" moments and sure enough it came true. I had been trying to get down with my weapon in my hand and the next thing I knew I had lost my balance and footing. Realizing that I was going to fall got me thinking about how I should land on the pavement below. Should I try to land on my feet like a cat? Should I land on my hands to brace my upper body and head and run the risk of breaking an arm or wrist? Should I try to roll like they teach the airborne students? What should I do with the gun? Well, I chose to throw the gun and roll. There was a time in my youth that if I had fallen with a beer in my hand that wouldn't have been an issue. At all costs I would not have spilled the beer regardless of what happened to me.

When I was in college a bunch of fraternity brothers and I would go to a swimming hole near my house. Cummins Falls was a great place to swim, hang out in the waterfall, and drink. It was on one of those adventures that I slipped on some creek moss with a beer in my hand. My face hit the rock with the force of a Mike Tyson upper cut. A gash opened above my left eye that definitely needed stitches but I didn't spill a drop of that damn beer. I rolled over and, to the amazement and applause of my fraternity brothers, got up and took another swig. The girls with us were aghast that I should be so careless about my head over a beer. Didn't they understand? Apparently they didn't. We stayed at the falls for the rest of the afternoon. I had a scar to remind me. You think I would have learned.

We unloaded all of our equipment and headed back to the CHUs where we found mail. It was good – Oreos, Snack Wells, movies. The people back home have been awesome with their generosity in sending stuff to the other soldiers and me. It was very humbling. I couldn't wait to get back home, do my road trip, and thank them all for their support of the American troops.

Later that afternoon we got slotted for another mission. We returned in time to catch Valentine's dinner. It was crab legs, shrimp, steak, and lobster. Awesome! I had two lobster tails, some shrimp, and a steak. My appetite had been down lately, which was a good thing, but I would have loved to have had a few more of those lobster tails. The food has been amazingly good here up to this point.

FEBRUARY 15

After rearranging more of the pictures on my computer, I headed to the computer center to check my email. This remained the biggest anticipatory excitement for me while in Iraq. There was always something good and I never knew when I'd hear from someone I hadn't spoken to in years. I also worked on catching up in my journal and picture library and then took some time out in the afternoon to hit the gym before dinner.

Later that night I was shanghaied as part of the 2100 patrol in Balad Ruz. Gunro would be our patrol leader on this trip. Big Red was finally allowing some of the squad leaders to lead patrols. We all needed to get used to the fact that Big Red and Dark Wing wouldn't be with us on every mission. Both of them had leave coming up. They wanted to give the E-6s some leadership training.

The squad on a whole and Gunro had not quite figured out how to interact positively with each other. For our part, we were used to Dark Wing and how he did things. Most of the guys were smart asses who did not listen to anybody but Dark Wing. Gunro did not know how to work around that. He was always degrading someone, most of the time guys on his own team. Most people responded to him by talking about him behind his back or lashing out at him with heated remarks. Something told me that his issues ran much deeper than he'd ever let any one of us know.

While Gunro was giving the Op Order he continually reminded us that he was an "E-6 doing an E-7's job just like when I was a Marine." We didn't know what the hell he was talking about. What that bit of information had to do with anything we did not know but it came back to bite him in the ass later that night. He ripped into Boo Boo for stepping away from the Op Order to take care of the radios. Gunro said, "It's all about respect." Boo Boo stared at him.

When we got into Balad Ruz, we dismounted in a different location than was given in the Op Order. We were a block away from where we were supposed to be. We all ragged Gunro amongst ourselves because he had been so proud of doing an E-7's job as an E-6. As I've always said, I like cocky but I hate arrogance. Being a block away was not necessarily a bad thing on this night because we were not going into a hostile situation, but the situation just added fuel to our fire in dealing with Gunro.

We ran into a few Hajjis who were out on the streets but most were old men just going home from friend's houses. They all looked tired. The moon was out, the night sky was clear, and the air was warmer than it had been in a long time. We carried NVGs at night, but on nights when the moon was out I didn't normally use them because my eyes worked pretty well in the dark and I liked using my natural night vision. Granted, there were times when I really needed to have the goggles on but I felt confident enough in our patrol at that time that I didn't utilize them. I would use my goggles for longer distance things. They were amazingly clear at a distance. That could give me the edge and save my ass on a night patrol.

We patrolled for a while, along some streets we'd been down a few times before. It was nice going through familiar areas because we noticed things that were out of the ordinary. In new places we didn't have any clue as to what to expect or what was out of place. We ended our patrol at the traffic circle. We had to wait on the Bradleys to come pick us up; we set up security in the market. Pretty and I held the twelve o'clock position on Route T1. The IP started to patrol east down Route T1. We were taking bets as to whether or not the IP would mistake us for bad guys and shoot at us. Fortunately they recognized Joe and passed us by.

February 16

We were slotted to break track on one of the Bradleys. It wasn't a hard process once we found a system for it, but it could still be a pain. We'd pull the Bradley up and bleed out the tension on both tracks. Then one team would drag new tracks into line with the old ones as two other teams would start the process of taking the old tracks apart. Once we broke the track we could join the new sections of track to the bottom sections of the old tracks. We'd lay out enough tracks so that when the entire track sections were on we could simply drive the Bradley forward and the new tracks would roll onto the wheels. Of course, ever so often we'd stop the Bradley long enough to dismantle an eight shoe section and load it onto a pallet so that it could be shipped back to wherever it was that the tracks came from and could be refurbished.

The hardest part of changing a track was reconnecting the first and last sections of the new track. We could join all the other sections in the amount of time it took us to join those last two sections. We had a good team that usually worked together to do tracks, but for some reason we always ran into something that gave us a hard time near the end. This time Big Red had come out to help us. It was encouraging to have an officer come out and help with what the enlisted guys were doing because most of the time officers came off with a separatist attitude. Big Red was not like that. He didn't fraternize. He would be in the mix just like any of us when the situation warranted it.

I finally was able to go back to chapel that night as we did not have anything scheduled. It was nice. I met Fog Head for the service. The service was good. The music was good and one of the songs sung by the worship team was "I Have a Maker," a song that we used to sing all the time at Young Life. It flooded my mind with memories of clubs, camping, and kids. It was beautiful!

When the time for prayer requests came up, I asked for prayer for Scooter and his wife. She had been put in the hospital this morning due to complications with her pregnancy. From what Scooter had told me, she could die during the delivery because of some blood disease she had. He was scared. I was filled with excitement because I had been given the opportunity to lift another person up in prayer and knew that a multitude was going to be petitioning God to intervene on his behalf. That was an awesome confirmation to the power of prayer.

February 18

Today we went into the market area to shop. Our mission was supposed to SP at 0700 but because of whom we were rolling out with, we didn't leave until after we all had breakfast. On the way to breakfast Mr. Murphy attacked our Bradley. It was in bad shape. The gear shift was not working properly. After chow we changed Bradleys for the adventure to the market.

Two reporters from the Army, one male and one female, rolled out with us. The guy was from Pittsburgh. He seemed to be a real good guy who fit in with us easily. The female reporter was a petite girl who had been a flight attendant prior to joining the Army. She was fairly attractive. We all made comments about her enough to where Big Red had to talk to us about holding our tongues in check while we did this patrol.

We rolled out of Caldwell and headed west on Route T1. We were to patrol an area near the Mayor's house first. We got rerouted to clear the western boundary. While we were there, Big Red allowed us to test fire our weapons. We set up some bottles and cans on the berm ahead of us and then opened fire. The female reporter got to fire the 50 caliber machine gun. Bong Water and Clementine fired the 25 mm cannon on their Bradley for the cameras. They acted like school boys who'd just lost their virginity to the most awesome girl in high school. It was orgasmic for them.

After shooting, we loaded up and headed into town. We parked along the market street, dismounted, and began our foot patrol. I took point. It was surreal because so many people were out, and in situations like that, one had to be constantly watching everything in one's sector. The problem was that so many stimuli would bombard us that it felt like we were always missing something. Of course we were, but fortunately, nothing happened on our patrol that we did not instigate.

We went down a few back streets and hit some clothing shops as well as some gold shops. I didn't buy anything. One kid did try to get me into his shop to play PlayStation. After an hour or so of our patrol we headed back to the vehicles to give the other guys an opportunity to go shopping. I stayed on the Bradley because CFB and I had started treating some of the scrapes and dings that the local kids had. When the next patrol was ready to go out, I told the LT that I would just stay there and help CFB.

The patrol headed back to the market and CFB decided to catch up with them. Crazy and I spent the rest of the time entertaining kids and trying to get bread. The kids always begged for something, whether it was for our gloves, our watches, or the scissors I had attached to my IBA for emergency purposes. For the first 15 minutes it could be cute but after 15, it got to be more than annoying. They crowded the back of the Bradley and we had to make them scatter a few times because they kept trying to come inside the vehicle. At one point three of the kids started fighting because they were pushing each other. We had a good laugh. Everything here was based on violence. It was what their culture taught them.

Big Red and Dark Wing working with IA

I ran into CFB and some other medics at chow. They invited me to play basketball with them at 1400. I changed into my PTs and headed to the court. Six of us played full court three on three. CFB and his roommate Lumberjack rode each other the whole time we were out. They were ruthlessly competitive with each other. CFB always seemed to get the best of his roomy; besides, Lumberjack was more of a whiner than a competitor. It was easy to get under his skin because he thought he was so damn good. But after playing with him that day I decided that I would not play with him again.

After the two-hour game, I came back for a promotion ceremony within our company. Only Boo Boo got promoted in our platoon. It was the fourth time he'd been promoted to Specialist. It really ticked me off. Some of these guys bragged about being busted and making their rank again, like they had gotten over on the system. To me, it just showed how shallow they were and how lame the system could be.

The ceremony was short but the meeting was too long. Our administration guy, OCD, answered questions at the end and some of the guys asked some really stupid questions. Don't let anybody ever tell you there aren't stupid questions. Finally, Dark Wing asked if chow would be over soon and if so, could we at least try to make it to chow. Everyone applauded, even OCD. We filed out and our squad got in gauntlet formation to "pin" on Boo Boo's rank.

February 19

We rolled out of Caldwell around 0900. Our mission was to go into Balad Ruz with the communications team, two spy guys with a drone, and their Captain, drop them off at the JCC, along with Priest, and then provide extra security for the JCC squad. A Muslim holiday that celebrated the lives of persons who had been killed the previous year was happening citywide. Balad Ruz supposedly had over one hundred thousand people living within its city limits. With the celebration, another twenty thousand were supposed to be present.

The city was blocked off at CP 1 and 2 so we headed south on Route C1 when we reached CP 1. Our plan was to skirt Balad Ruz as much as possible and then enter the city from the south. This would allow us to steer clear of the celebration while maintaining good security in the area. We accomplished this by patrolling through the rice paddies and fields south of the city. Kids were out everywhere and waved at us with great excitement as we passed their homes. It always made me question why the insurgency never did anything on our back routes. We went down them enough that our routine should have been easy to distinguish.

For the celebration, the Iraqi National Guard (which was finally being called the Iraqi Army) had beefed up its presence by having another company of soldiers stationed throughout the city and surrounding country side. This was beneficial to us because it freed us from the duties of doing checkpoints. It also gave the IA an opportunity to take ownership of their operations and country. For the most part, the IA was really good at what they did. Sometimes though, when they set up checkpoints, they were not completely fortified. They were much easier targets for the insurgency assaults and ambushes. The insurgency went after American forces with roadside bombs.

At the JCC every one dismounted and took up positions either in the courtyard or on the mayor's office as roof watch. We saw throngs of people to our west. They were at a civic building for the memorial service. I suspected they did the service there so as not to offend either sect, Sunni or Shiite. When the service was over, the people marched down the main street. Groups were led by standard bearers; death banners and religious flags escorted both the grieving and celebrating locals. When they reached their respective streets, they disbanded to the areas they had previously designated as their rally points.

After the procession a bunch of IA came to the JCC for lunch. Most of the guys were from Charlie Company so we recognized them. They were a good bunch of guys to get to know and work alongside. Major Ali also made an appearance. It was good seeing him as well since we were told that he was getting a promotion which would move him out of his job within the city and into Caldwell to train new recruits. He would be sorely missed in the city as he was the force behind ridding the city of bad guys. He was very aggressive in his position and with his style of leadership.

Dark Wing and Little D bartered with some of the IA. Dark Wing walked away with some pictures of Saddam Hussein which he had gotten in a trade for a flash light. Of all the things the IA wanted, flash lights, gloves, and knives were at the top of the list. Most of the time we had the stuff to trade, but for some reason on this day I had nothing to trade so I just took pictures.

We mounted up after the IA left and headed out of Balad Ruz with one Iraqi who had mistakenly been arrested. I manned the 50 for the ride home. We took a different route which went through some of the south side of the city limits and palm grove. Again, kids and families were everywhere asking us for things as we passed them. Some kids simply waved at us with their thumbs up screaming, "Mister! Mister!"

Boo Boo and Atlas started a shelf-building project in our CHU so I got out of the area and went to the computers to check my email. Atlas had been a carpenter in the States for some time. He was still very efficient at it. They were able to install five shelves, three for Boo Boo and two for me. With limited space, the only way we could go was up.

At 2100 we had a squad meeting. Crowfoot decided to pass information like this so that he would not have to chase people down and repeat the same thing over and over. I hated meetings but fortunately this one was not long and was very "to the point."

FEBRUARY 20

In the morning I worked on my journal for a bit and then went to the computer center to check emails and send off my jersey logo and shirt design to AmericanRugby.com. At some point I'd get their response with costs and estimates and then we'd be able to have Deuce rugby shirts for our squad. With a little help from Dark Wing, Boo Boo, Clem, and Crowfoot, I designed a shirt with blue and white opposing quarters on the body, black and white stripes on the sleeves, and yellow collars and cuffs. The Deuce logo would go on the left chest quarter, the white field. Black numbers would be on the back on white, diamond fields. Our names, in black, would also be on the back.

After a light lunch I came back to the CHU to relax a bit and read more of *The Godfather*. We got mail so the first bits of my reading were notes from home. I had also gotten a few packages. With the mail read, I dove back into my book. I looked forward to seeing the movies when I got back home.

I went to the gym at 1530 to get in a quick workout. It was a good burn. The whole time I was there I thought about Special Forces. Should I really go for it? Was I too old? Was life finally catching up with me? Could my back make it physically? I thought I could do it but I wrestled with my heart as to whether or not I really wanted to pursue it. I wanted to operate at another level. There was something to be said for working among individuals and with a team that would challenge me to be the best I could be every time I was in their presence.

From what I'd experienced with special groups, most of them could care less about rank although they did respect the positions of leadership within the team. They wanted to be known for their efficiency in job performance and reliability among their teammates. In a crisis, whom could I count on and who could count on me? I wanted to be a part of that social order – silent, professional order.

There was nothing worse than working with people who didn't want to be at a job or do their best everyday. Granted, there were times when I did not want to do the job but that wasn't the attitude I had day in and day out. Other guys did not want to be here and their attitudes reflected it. Why were they here? Why did they volunteer in the first place? Was it just something to put on a resumé? Was it something they could do one weekend a month to earn some extra cash? Nothing wrong with that but they had signed up for the military. Combat was expected. All I wanted to do was be the best at whatever job I was assigned and move forward with a professional team.

After working out and catching chow, we all headed to the track line for our patrol. We were going to patrol the market area again. We rolled out at 1900 and dismounted near the hospital. Our patrol route was one that we'd done several times before. Since curfew had changed more people were out. It gave us a chance to interact with them in a positive light. At one point, though, as we were passing a corner, two guys took off running. We gave chase. One guy got away from us but we did locate the other one. Big Red brought him out of his house for questioning. His wife was a teacher who spoke some English. She was able to translate for us. He did not give us any useful information as to where his buddy went, but we did get our point across that he shouldn't run from us when we approach him because it makes us think he's a bad guy.

A lot of the men who were out were drunk. This was strange because alcohol was not supposed to be present in the area. It was driven down our throats that while we were in Iraq, there was no alcohol or pornography allowed because of the Muslim religion. We weren't there to offend the people of Iraq. We were there just to kill them if we thought they might pose a threat to us. It was funny because on any given patrol some man or boy would ask us if we wanted whiskey. We always had a good laugh at that especially since it was part of the Big Red One's General Order Number One.

When our patrol was through, we mounted back up, headed to the JCC, offloaded some water, picked up one soldier, and then headed back to base. A few of us went to midrats and caught some of the Daytona 500 being broadcast in the chow hall.

****

From: *<Larry>*
To: *<Malcolm><Mike><Mom>*
Sent: *Sunday, February 20, 2005*

Sort of looks like the Disneyland gang caught out in the desert! Donald Duck, Mickey, Pluto, Goofy …

But I want to thank you belatedly for sending us the very well written (!) account of your recent patrols. It was intensely sobering to read. It was chilling. I shared it with Dennis, our family friend who was injured in Vietnam. He read it so carefully. I thought he'd never finish. I've read it several times and thought about it a lot. Your descriptive power is as good as any war correspondent around. And what you describe is true grace under pressure, as John Kennedy would say. Ironically, as a sole survivor of someone killed in action, I was spared from serving in Vietnam or the terrible dilemmas of opposition. Challenges that uniquely forged character in my generation. And I regret that. In a way, you guys are climbing the mountain that we here only see in the distance. Your email has been the telescope.

February 21

We weren't scheduled for anything until 0950 so I spent the morning working on my journal. At some point I was going to have to catch up on my living space as it was getting a little cramped with the piles I so enjoyed making. It was almost like the filing system I have on my computer. Only since it was out in the open, it wasn't quite as pretty.

We began our patrol at 1000. Our plan was to go into town and find an Iraqi man who was selling anti-American propaganda. When we had been on our market patrol a few days earlier, Big Red had picked up a couple of CDs from the guy and when he watched them he saw images that encouraged the Iraqi people to rise up against American forces. In particular, there were segments glorifying the Jessica Lynch incident as if the particular group who had made the video was responsible for that convoy ambush.

Big Red really wanted this guy. He had ordered our entire platoon to go on this mission. He had also arranged for a platoon of ING to assist us with crowd control. Crowfoot, Crazy, Pretty, and I were the detainee control team. We would not actually do the snatch and grab, but when the detainee was restrained, we would jump out of the Bradley, take custody of him, and quickly get him into the Bradley where we would ensure his "safe" travel to the detainee center on base.

Fortunately for him he wasn't on the streets that day. We did, however, run into another video salesman but he was only selling Muslim holiday videos. We surrounded his area as coolly as if nothing were going on and had him show a few of the videos. After we were content that he wasn't selling anti-American propaganda, we mounted and headed to the JCC where we got to see Fred again. Fred was the JCC dog. He was a dirty, mangy mutt but he followed squads around on their patrols if they left from the JCC. He was fun to have around and would act as an early warning system on those patrols.

After everything was taken care of at the JCC and with the ING, Big Red came out and told us that our mission had changed. We were now going to a location in our AO that none of our troops had ever been to and check it out. It was an hour drive from where we were. Since we only had four guys in the back of the Bradley, we could spread out and catch a combat nap.

We stopped at two little towns, if one could even call them that. There were a few houses together like a compound. We checked out the first cluster from a distance. At the second one, we dismounted and snooped around. The people were very friendly once Big Red conveyed to them that we were only there to ensure our and their safety. One gentleman led us around the compound. He wanted to make sure we weren't disturbing the women too much or the wedding ceremony that was going on in one of the houses.

While we were searching, we found an AK. It was no big deal other than Crazy found it in what appeared to be an outhouse. Not only did he find it in the shack but he dug it out of the hole where a water pipe came out. It was funny but nasty. As I said before, Iraqi families were allowed one weapon. We had to reassure the man that having that weapon was all right and we weren't going to do anything to him or it. He seemed quite relieved once he understood what we were trying to convey to him.

The compound had a lot of spent artillery shells from wars past but nothing illegal. The shells were being used for table stands in the courtyards. We gave out candy to a few of the kids to calm them down from the scare of having strange men coming into their homes with guns. After taking a few pictures of them and letting the kids see the pictures, they began to trust that we weren't there to hurt them.

We were satisfied that nothing was out of the ordinary there. I took one last group photo and then we mounted up to head back to Caldwell. On the way back the Bradley's hydraulic engine for the ramp door was running wildly and the engine was overheating. Fortunately, we made it back to the motor pool before anything really bad happened, though throughout the entire return trip the four of us in the back of the Bradley were planning our escape routes should anything start burning. We weren't going to be trapped inside that mobile tomb.

FEBRUARY 22

Today was the first day of our green cycle. We hadn't planned on doing too much so as I got up at 0800 I thought I'd have a full day to get some things done that I'd put off like writing letters, cleaning my living area, and taking my uniforms in to be laundered.

I spent an hour or so answering emails and then headed to the PX to see if any new items were on the shelf. There were a few things there that I wanted so I bought a new M9 leg holster. I had a holster issued to me but the system was archaic. When I say "the system" I mean a webbing belt and a holster which had to be linked together to be used. The new holster I bought was a one piece drop-down holster I could use off the belt I already wore with my DCU bottoms. I hadn't been taking the pistol out, but now I could start taking it on patrols with me.

Word came to Bong Water today that his sister was brain dead in a hospital back home. The family planned to pull the plug on her life support the following day. He was a little upset and was trying to find a way to get back home with the help of the Red Cross. He had been the lead pall bearer at his mother's funeral and expected to carry his sister's ashes in her memorial service. He was able to leave the next day.

At 1300 on maintenance days we usually worked on our vehicles. Today was a little different. We were told by Dark Wing to meet at the vehicles and that we'd be going to the flight deck to learn a little bit about mounting / dismounting from Black Hawks. We were slotted for an upcoming mission with ODA. Our part in the mission was the air assault. We weren't told anything about the mission. The only thing we knew was that we needed to be proficient in loading, riding, unloading, and pulling security, which meant we got a safety class taught by one of the crew chiefs.

With the class over we headed back to our vehicles. As we were leaving an ODA fire team was brought in on another chopper. They busted hump while they were there, but man, did they look like they had it made! They did top secret missions in their street clothes! I loved it. I found myself, once again, wanting to be one of them, doing clandestine ops, riding choppers, dressing down, and training and fighting like soldiers were supposed to train and fight.

I went to the gym and did some push-ups and sit-ups. I then loaded my ruck (my backpack) and went for a two mile ruck march. It was good. To be a part of Special Forces, or to complete any significant school in the Army, one has to do ruck marches so my plan was to get my body, particularly my legs and feet, used to doing them. They're somewhat like patrols except in a ruck march soldiers don't stop and talk to the children in the area. I threw 40 extra pounds into my pack and began my march. My legs were burning and the straps from my pack were cutting in after the first half mile, but it was a good ruck nonetheless. It was the first one I'd done since medic school.

After cleaning up I headed to evening chow. My evening was slotted for tattooing. Pretty had finally sacked up and given his word to get the "Don't Tread On Me" snake and words tattooed on his upper back. At 2000 he showed up and a room full of us began the harassment. He did well for his first tattoo as did everyone else in the CHU for their part in harassing him.

FEBRUARY 23

Today was another green cycle day – not too much going on at all. We had taken care of all our maintenance the day before so none was scheduled for today. One of the highlights was running into a couple of medic buddies from 3rd Squadron, TC and Lineman. They had stopped in Caldwell on their way home for some R & R. They had already been on post for a couple of days and were just waiting for their ride out to Anaconda. Lineman wanted to work out so I hooked up with him at the Regiment gym at 1330. We worked out for a little while and then he left to get ready to move on to the Air Force base.

That night I played Scrabble with Dark Wing. We split two games. This was the first time Dark Wing and I had done anything on our own. DW was always running with the other guys. We just didn't hang out much even though he was only a year or two older than I and we were both prior service. I guessed it was my personality. He liked the youth and mischief that the other guys offered. He seemed to be an intelligent guy, always reading, and I thought he'd be a good challenge on the Scrabble board.

After our games I headed back over to the CHU to pack up for the 0800 dry run at the Black Hawks. We were told that we could possibly be spending a couple of days out in the field on this mission. We packed as lightly as we thought we could but still had a bunch of items that weighed us down, especially in Boo Boo's case. For weeks we had been telling him he needed to offload some of his equipment because his pack was too heavy, but he would never listen. The next day he would be learning from the masters.

FEBRUARY 24

At 0800 we met at the vehicles. After loading enough equipment for a few days in the field we headed to the helo pad. Once there we learned what we needed to do to on-load and offload as quickly as possible for our upcoming mission. The ODA guys were there as well to do their dry runs. Our platoon was split into two teams with Black Water Six leading one crew and Big Red leading the other. I was in BW6's team.

We ran through the drill five times. With each run we attempted to reduce our on-load and offload times. We had assigned seats on the bird that we'd be flying on to make the loading phase quicker and more routine. We would board, hook up our seat belts, and then pull our assault packs onto our laps. After we gave the thumbs up that we were locked in, we would then practice getting off the bird, first throwing our gear on the ground and then jumping out and setting a perimeter.

While we were at the air strip, a couple of Apaches were refueling. They were bad, wickedly intimidating. The Apaches would be our air support during the mission. It would be great to be an Apache pilot but it sure would suck to be an opposing force.

I talked to the ODA Doc, asking him what essentials he thought I should bring on this particular mission. He said, "Just bring enough shit to patch holes. All you want to do is stop the bleeding 'cause we're not going to be that far away from the birds. Fuck calling in the medevac. We'll just throw them on the birds already there if something goes down. Don't worry about bringing a lot of fluids. We're going in and dominating the place but just be ready to patch holes. I'll bring my BFM [his medic bag] but once we get the place secure I'll put it down in a place where I don't think anyone will steal it." He was the pro, so when I got back to my CHU I offloaded quite a bit of stuff that I didn't really need.

Later that night, we had an informal platoon meeting where we were blasted because someone had broken Op Sec (Operational Security). A reporter from the *Chattanooga Times* had asked BW6 if he could go on our upcoming mission. He said that one of the guys from Apache had informed him about it. Needless to say, this infuriated BW6 and he threatened to take away all our privileges, even our mail, if the person who talked didn't confess. Big Red and Dark Wing went to bat for us and our privileges were not taken. The truth came out about the breech of Op Sec too. BW6 forgot that he had mentioned it to the reporter. We never got an apology.

****

From: *<Malcolm>*
To: *<Mom>*
Sent: *Thursday, February 24, 2005*

I've enjoyed communicating with Peyton too. She's got a sweet heart. I know I can count on her to take care of the guys here, especially the guys who don't receive much if anything from back home. That's really cool of her. Love, love, love. That's what it is.

FEBRUARY 25

Our mission started at 0900. We met at the briefing building and had a equipment and personnel inspection with Dark Wing. By 0945 we had loaded onto an LMTV and were headed out to the air strip. Because of the timing of the mission we were kept out of sight in a holding area between some buildings. We didn't want any Hajji that was working on base to know what our intentions were in case he thought he might make a run down south and inform Ali Baba that we were coming to get him. The ODA guys joined us and we all lay around waiting for the mission to begin.

The operations sergeant for the ODA team told us that the window of opportunity to snag our target would most likely be between 1100 and 1200. Supposedly, Ali Baba would be meeting with his top aids during this time. They had an informant on the ground who would call in to give us the go ahead to move on the subject. We were to expect heated resistance. ODA believed the compound was using an advanced warning system to alert Ali Baba of any coalition movement. Supposedly there was also a weapons system that could tear through Hummers somewhere in the city. We were going in somewhat blind.

Since we were in the holding area so early, most guys crashed out. A few of our storytellers began swapping tales about their pasts with one another. The ODA Doc brought out a case of Gatorade and a case of Sprite for us, compliments of their team. I thought that was pretty cool so I broke out my beef jerky stash and shared it with as many of the guys who could get to it before it ran out.

Conversations ranged from how one met one's wife to what Joe would be willing to do for a million dollars. We also hit the events involved in one's first arrest. Most all of the stories brought a good deal of laughter. Because most of our soldiers (and soldiers in general) had a great deal of machismo, we often found ourselves in situations where we felt like we had to "one up" the other guy. The stories got better and better as the day went until finally everyone got bored with the outlandishness of them.

The mission never happened. The informant who was supposed to call in the target's presence only called to say that the target had never showed. Everyone returned to his living area and stayed at the ready just in case he called back. By dinner, we were in complete stand down. The mission was not going to happen this day.

February 26

Some of our platoon had a 0400 mission this morning. I wasn't slotted to go so I slept in. It was nice. The days had started to get longer and warmer and the nights shorter and much more enjoyable. Boo Boo and I left the HVAC system off overnight and it got pretty chilly in the CHU by the morning, but it felt good.

After I woke up I fiddled around on the computers and then came back to the CHU to square some other things away. A few of the guys had wanted some of the music I have on my computer so I burned some Metallica for them as well as downloaded Stevie Ray Vaughn into my collection. I did finally get to write my friend Mason Jar after a week of neglecting to do it. She had been so good to write me throughout my whole stint away from home – ever since I left for Texas to do medic training. She was a sweetheart.

Our SP time was 2300 and we were slotted to set up an ambush in the proximity of an old IED site. As fate would have it, we didn't do that mission. Instead, we did a foot patrol in the suspected area of a mortar launching from two nights before. I walked point with Percolator behind me on the SAW. We headed south from Route T1 and planned to link back up with the vehicles at a road a few clicks south of Balad Ruz proper and then head out to Route C1 to return to the base.

Walking point continued to be a rush for me when allowed to do it. Every sound, movement, and instinct had to be taken to heart. The second I decided not to trust my instincts or rely on my training could be the second somebody put a round through my throat or I hit a trip wire. No point man wanted that to happen or wanted to lead his patrol into an ambush. Fortunately none of these things happened on my watch.

The thing about walking point, besides the fact that one is on a higher alert than when not on point and that one had to know what was going on ahead, was that one had to be alert for anything behind as well. Sure, there were times when a soldier might actually miss something or choose to let something slide so the guys behind him could check it out, but if something went down, he had to react to that while taking security on his end. It was a great experience that not everybody volunteered to do, but I thought it heightened my experience. It made me a better soldier.

The foot patrol went off without a hitch, but five minutes after we mounted the vehicles we had to stop to pull barbed wire out of the track of the Bradley. The little roads we patrolled were full of hidden threats to a big vehicle but surprisingly we'd only hit wire twice. Our guys had the wire out of the track in less than ten minutes. All of us who were pulling security got back in the vehicles and we headed out. We thought we were going straight back to base but we ended up skirting the city of Balad Ruz first and then we headed back.

February 27

We just got back from a night patrol. I walked point again through the city of Balad Ruz. What is it about walking point that I enjoy? I think it's the fact that if anything goes down I would be the first to see it if it's in front of us. Obviously, if something happens to my rear I have to react in that direction and more than likely will be behind the curve for a few seconds; but in the front, I'm right there.

We ran into all sorts of things while on patrol. Most often we ran across Iraqi citizens making their way home or hanging out with friends if it was at night. When a car approached us, we would first try to get the driver to cut his lights a good distance from us so that we could use our flashlights to our advantage. We would shine a light into the cab, blinding the driver and passengers for a brief time. Next, we would check out the car. The Army called this a hasty check point. We called it a Roach Motel, and when we did it we would get the driver and all passengers out of the vehicle, pat them down, and then thoroughly inspect the vehicle for any weapons or hostile material. Most of those checks had turned up very little.

Our patrol tonight turned up nothing out of the ordinary. Black Water 6 had changed our mission a little bit because one of the city council members had been threatened. We drove to the councilman's house and checked on him while we were out just to make sure nothing was happening to him or his family.

On our return trip, a few hundred meters from the base gate, we were passed by two vehicles, the rear one driving without its lights on. Big Red ordered up a Roach Motel and we held the vehicles in check with the Bradley in front and the two Hummers in the back. Since the Bradley boys were so far in front we didn't approach the vehicles. After clearing the occupants we mounted back up and headed back on to the base.

Tonight, Pretty and Little D were rendezvousing with a couple of base women for what they believed would be a long awaited, sexual encounter. The gals were from one of the supply companies. They had met at the chow hall and had set up the meeting for when they got back from our patrol. We'd see what kind of game these guys really had considering they always talked about girls and getting laid. Little D set his standards low so he wouldn't be disappointed. It was certain that he, at least, would score.

Beware the Ides of March

March 6

We arrived at the JCC on the 2nd and were immediately told that the ING was in the process of setting up ambushes because of some Intel they had received from a local who had been approached by the AIF. Apparently the AIF were planning to lob mortars in our direction and told the guy to get out of his house so that they could use it as their staging area. Fortunately for us, the guy rolled over and gave us the information. We were ready for anything they were going to throw at us but good fortune and God were on our side. Nothing happened.

This rotation was the first one that Shrek was with us as Sergeant of the Guard. He was a prior service guy and had been a cop in Knox County for seven years prior to this deployment. He liked to remind us of that. Not many of his own troops liked him in the position he had been placed. He had what one would call "an ass kisser's disposition." He always spoke his mind whether anyone wanted to hear him or not. That got annoying. On his behalf, he wasn't too bad one on one. It was when he took on leadership roles that he became a pain because he got too excited.

He spent the entire time we were at the JCC arguing with people. We had been running the show fine for three and a half months without him. He did not need to come in and shake things up for his own edification. He was an E-6 and his rank deserved the respect from everyone below him; but again, he was a poor leader and all the guys felt that way about him, even the guys more level-headed than I. I stayed clear of him as much as possible and tried to do everything he asked as quickly and efficiently as possible so I would not have to listen to him more than I had to.

For some odd reason, there was a lot of tension this time at the JCC. I did not know that it was not so much because of Shrek's being there as it was that everyone in our platoon was getting tired. We'd had over half of our platoon on leave for more than two weeks and the fifteen or so of us who were left had been covering all the missions and JCC duty.

I was getting tired of doing police work when I had volunteered to go to combat to fight an aggressive war on terrorism. Most of our patrols were spent walking around our AO looking for evidence that bad guys had been there and while we were out we were just targets of opportunity for anyone who wanted to take a shot at us, be that a bullet or a bomb. It was unnerving for me because that was the one thing I did not want to happen while we were here. I did not want us just to occupy their country and train their Army. I wanted us to fight the War on Terrorism. Most of the time I felt like all we were doing was acting as a non-aggressive police force.

We could get about anything we wanted while at the JCC. We had guys getting AK-47 bayonets, food, liquor, and unfortunately, some drugs. Drugs and alcohol I could deal with, but when a soldier did them on watch it was a totally different story. Two of our guys had done that so far while we'd been at the JCC. Worm took Valium this time and got so wired he could not finish his watches. This infuriated a lot of us because it was soldiers who acted like that who would end up getting other soldiers killed. We also had another guy fall asleep on watch which, if left unchecked or uncorrected, could result in the same thing. The incidents were addressed by Dark Wing when we got back to the base. I said something to Dark Wing about Worm because I didn't want guys going out like that while I was on patrol. He needed help.

Other than those isolated incidents the rotation went down without any major problems. We did our duty and on Friday morning we were picked up by 1st Platoon and brought back to base. I manned the Mark 19 grenade launcher on the way back. A dust storm was blowing through and I did not have much face protection so the sand hit me like little darts. It was definitely an interesting ride back trying to shield myself in such a way as not to come back looking like I'd been stung by a swarm of bees.

We were slotted to retry our mission with the Special Forces guys. Because the weather was so bad, brass had decided to make the run in Hummers rather than in Black Hawks. We were all bummed about the mission because none of us wanted to pass up the chance to do an air assault, but with things as they were, we knew that a flight was out of the question. We staged at our CHUs and waited for the word to move out. It never came. Again, we never got the word from the informant to move - another mission scrubbed.

Our night mission left the base at 2300. We headed south again. We hadn't been south in a while and this had been from where most of the activity in our area was coming. We rolled out in a Bradley and two Hummers. Those of us in the back of the Bradley couldn't see much as we rolled. As we approached our staging area, we heard the guns firing. When we heard the guns, excitement grabbed us.

The recoil of the guns shocked our systems and immediately we were all alert, wondering what was going on. Crazy got on the internal comms to talk to the Brad crew to find out what was happening. Apparently, some suspicious individuals had been out after curfew in a field. As we approached their location, they started to run. Normally we wouldn't give this kind of activity a second thought in America, at least not unless we were cops, but in Iraq a soldier had to immediately question what the intentions of these people were. Were they spotting for IED placement? Were they snipers? Were they just farmers checking their aqueducts?

Big Red made the call to pin the guys down so our gunner opened up on the area. He was a good gunner and was able to fire with precision and not kill anybody. Since the route to the area was not passable for us, we stayed on Route C1. Thermals were a great advantage for us. They could pick up anything that moved and could do it at a great distance. The range and accuracy of the guns was scary. The gunner and BC scanned the area. They stood ready to fire if anything more threatening or suspicious reared its head. The crews in the Hummers were able to navigate where we couldn't. They went as far as they could to investigate but the dismounts had to get out and do a patrol on foot the rest of the way. From where they dismounted they were not able to cross the berms and aqueducts in front of them.

It was then that Big Red decided to call in a fire mission. We were approximately 15 miles southwest of Caldwell. We heard the sound of the rounds being fired and as we waited we wondered how long before the whole area would be lit up. I had never experienced an illumination round going off above me so I didn't know what to expect. They didn't sound loud at all in the movies; they just looked real peaceful floating down out of the night sky. Well, that wasn't reality. The round separated not too far from us, but when it did, it startled most of us. Little D, on the other hand, was on the deck like he was about to get hit by a low flying airplane. It was hilarious. We all had a good laugh at his expense. Of course, he didn't think it was that funny; but after twenty minutes he was laughing about it as well. After the round burned itself out, the Hummer teams mounted their respective vehicles and returned to their position.

We headed back to base thinking that we could come back the next morning and see the area more clearly. Saturday morning we took off around 0630 and headed south again in only Hummers this time. The roads were slick from the previous day and night's rain. The ground we were going to be covering was thick with mud. Once again, I drove. After returning to the area where we had fired on the previous night and finding that the guys had in fact been farmers checking their aqueducts, we patrolled another area south of the city where the bad guys usually ran. This proved to be quite a feat of driving prowess.

This adventure was the most intense driving event to date, even more so than the big chase. We rode in Pucker Factor #10 the entire time. We were in four-wheel high, spitting mud on our gunners and on the vehicles behind us. The roads beside the aqueducts were dangerous and the Hummers slid back and forth over the rain-soaked earth. All that kept running through my head as I didn't want to cause the death of three other Joes was the old Simon and Garfunkel song, "Slip Slidin' Away." I ground my teeth and just kept singing that over and over to myself as we got within inches of going into aqueducts time and time again.

What normally would have taken us twenty minutes to travel took over an hour. I was the lead vehicle and had to make sure every spot was passable. There were some crazy spots. Just before we started heading into Balad Ruz, we ran across one of the Iraqis who worked at the Mayor's office. He lived in the middle of the field we were passing through and couldn't get to work in his car because of the depth of the mud. He was an interesting man named Shaker. We picked him up.

Anti-Iraqi Forces had attempted to kill him three times already, but he insisted on using no body guards. He invited us to eat with him anytime we got a chance. Because of his kindness I decided to give him a little treat in the mud. I fishtailed a few times and took us as close to the aqueducts as possible. He seemed to enjoy it, though he didn't look very comfortable. This part of the ride was a good break in an otherwise stressful day.

We took a small break at the JCC to clean the windows on the vehicles and snap some group photos. Fred was there so he got a good rub down from us. We mounted up and got onto the main street, heading back to base. We made it back without incident and went to the refueling station. As we made the turn into the area we saw another Hummer that was covered in mud. The driver and passenger had big smiles on their faces. We knew something was up. That something was huge mud puddles along the route to refuel; so being the boys that we were, we hit every one of them. Mud and water were everywhere and the more that got in and on our Hummer, the more we looked forward to getting it even dirtier. It was awesome!

The afternoon was pretty low key as we were getting ready to go out on another patrol that night. I was able to do a tattoo on Lug Nuts, the PA for 3rd Squadron. He had been my boss when I was with HHT – 3/278 ACR. I hadn't seen him since Kuwait. He'd gotten a tattoo in Shelby and needed it re-colored. I did it and then we went to lunch. He always had a good smile and laugh to offer a brother.

The mission schedule had us reporting to the motor pool by 1950. We made our way there but weren't too happy because Black

Water 6 was going on the mission with us. He'd been trying everything he could to be out enough so that he might be engaged at some point. Once engaged, unless he got killed, he would be able to submit the paperwork to be awarded a Combat Infantry Badge. He was the top brass of our company so he could pick and choose any patrol he wanted his crew to go on.

From a leadership perspective, he had a bad habit of taking things over from his Platoon Leaders and micromanaging. We could tell that it frustrated Big Red to no end. Hell, it was killing us that we were even going on patrol with him. It wasn't that BW6 was such a bad guy; he just did not come across as someone who exemplified selfless service in our eyes. He, like so many others, led to glorify and edify himself. That was scary for anybody below him because we were the ones who would pay for any bad judgment.

As we were gathering at the Bradley and telling the same old stories that had been told all day, Shrek came down and told us that the mission had been cancelled because one of our other teammates had a negligent discharge. A negligent discharge meant that someone had not cleared his weapon properly and his weapon had fired. The question we all had was why was it loaded in the first place? One of our routines was to clear the weapons every time we came back onto the base after a mission. To ensure that was done properly two guys checked the chamber and magazines of each weapon that came back on base. It should have been cleared but for some reason his weapon was loaded and he fired it. The bullet almost hit Cow Tits. Of course, this brought down the thunder. The mission was scrubbed and an investigation began. Everyone involved was questioned, even Big Red and Dark Wing, who were not in the area at the time of the shooting.

That was a tough situation because some of more immature guys in our platoon made the situation worse than it should have been. The guys were already having issues with Atlas when his weapon discharged. That incident just gave them more fuel for their fire. What they seemed to be neglecting was the fact that a ND incident could have happened to any one of us. We often took our routines for granted. Routine was an easy way for complacency to set in. Complacency got people injured or killed in combat situations. Fortunately, nothing happened to Cow Tits or anyone else this time but it was damn close.

Dark Wing having a stab session with the guys

I got up early to hit the computer room and get some chow before our 0900 platoon meeting to discuss the complacency that had overtaken us the previous week. It was a good meeting but I was not sure if Dark Wing hit things hard enough to put the fear of God back into us. I hoped so. None of us needed to take anything for granted over here, especially when it came to weapons and the mishandling of them with live rounds. The rest of the morning was spent working on the Bradleys. We did maintenance on them and cleaned the weapons systems.

****

From: <Malcolm>
To: <Becky>
Sent: Saturday, March 05, 2005

Unfortunately, with our mission schedule we don't get to do a lot of the base things that the soldiers who stay on base all the time get to enjoy. Usually we'll be on mission and when we're not we're either emailing, working out, eating, or sleeping. I do the occasional tattoo and another guy cuts hair but most of us just stay away from the rest of the base folk simply because we value our time off and don't want to waste it.

*Bummer about only getting four people in [for haircuts]. I'll probably do that many tattoos today but that will be good for me. I'll be wishin*g for a break after the second one.

MARCH 7

Today was day two of green cycle. I did a couple of tattoos. The first was on Percolator. I tattooed the words "life" and "death" on his forearms in Arabic. He had gotten the translated words from an Iraqi he had been talking to while we were at the JCC. It looked good as far as simplicity was concerned. I also retouched the "life" and "death" in English on his wrists.

Percolator was an interesting character, twenty years old and already with a lifetime of stories to tell. He had been arrested numerous times for violations involving drugs and alcohol. He wasn't supposed to be on this deployment because he'd popped positive on a whiz quiz that the Guard had given him the previous year. Just like most young people, he was always out to prove he was the man, tougher and crazier than anybody else in the crowd. He was just as lost as most every other kid his age. Life had put him in some situations and he had not made good choices. He did have a good heart though and one could not help but laugh with him about things. It was sad that his actions did not always go the same way his heart wanted him to go.

After lunch I tattooed one of the KBR workers from North Carolina, Mike. He'd been a Ranger in the Army thirty years ago. His tattoo was letters in the shape of a circle. OIF was inked in the middle with the letters ESA and NEM circling on top and bottom respectively. He said that only he and the music legend, Stevie Nicks, knew what the letters were about. He wouldn't tell me or his buddy, Jason, what they meant.

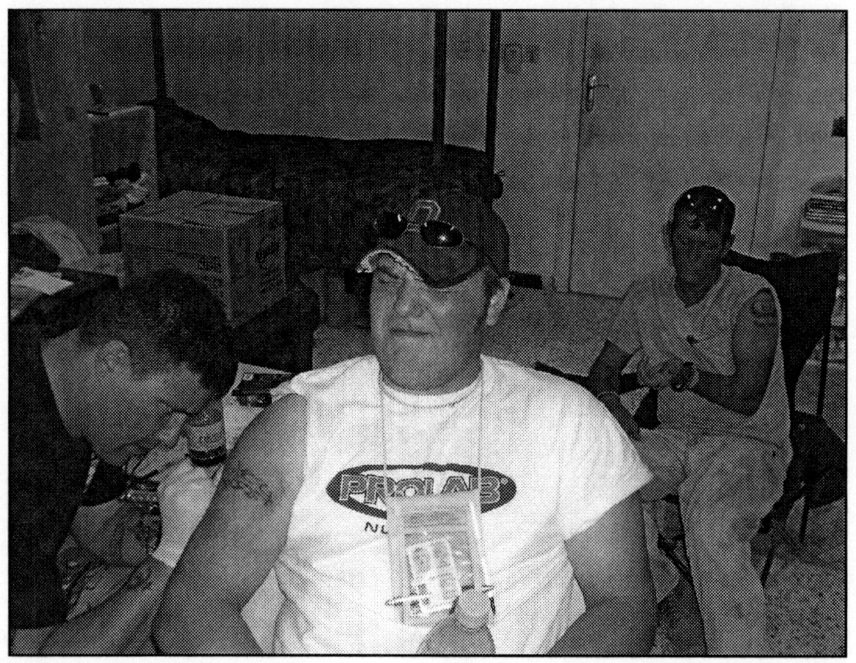

Contractor getting inked

At 1500 we had a company formation in the Chapel. Black Water 6 wanted to talk to all of us about the alcohol and drug use that seemed to be rampant in his company. It was a fairly good meeting but ended up being a little long winded. Because of that I think the talk lost some of its value. Nonetheless, the CO put his foot down and told everyone that he had wanted to do a drug test that day but the First Sergeant had persuaded him not to. He was going to hold off for a week. He did give the opportunity for those soldiers who had been using to be man enough to tell their chain of command during this week because if they popped positive afterwards he was going to do all in his power to reduce the soldier in rank, take as much money as he could, and dishonorably discharge the soldier after his tour of duty in Iraq. From what the First Sergeant said, a few soldiers came up to him after the meeting and confessed to what they had been doing.

The meeting lasted about an hour so that left us with thirty minutes until chow. I went back to the CHU and conned little hung-over Boo Boo to go to dinner with me. Boo Boo had left the CHU last night after watching a movie and I didn't see him until the next afternoon. Apparently, Dark Wing and Crazy had some Jagermeister and they got him hammered. They took a video of him lying on the floor, screaming after being beaten down by Crazy. He was out of control. At the chow hall we got a great meal of steak and lobster. It was nice and made better by the three lobster tail hookup by one of the servers. It was a shame that because of the condition Boo Boo was in, he couldn't enjoy the meal.

After chow I went to the First Sergeant's room to give him a tattoo. His was an upper arm band with a dragon head as the center piece. We worked on the piece for a few hours until my gun broke. One of the pieces that held the needle came off. I assumed it was from the heat or the looseness but it was the first time that had ever happened to one of my guns. As for the event itself, well, let's just say that the First Sergeant didn't like the pain too much. He was a good sport but he made sure that the next time we got together we would have some kind of topical numbing agent to put on his arm.

March 8

I wasn't slotted for the morning patrol that dropped off the JCC squad so I slept in. The First Sergeant had a morning meeting so we planned to meet at 1100 to finish his tattoo. On my way back from the computer room I ran into CFB and he hooked me up with some Lidocaine. At 1100 Top picked me up and we headed back over to his room to finish the ink work. It took three more hours to finish.

Doing tattoos for an extended period of time had a tendency to wear me out, especially when I was slouched over the client. I was bushed so I lay down for a while and watched an episode of *The*

Andy Griffith Show. I'd always enjoyed that show but watching it in Iraq made me really appreciate the simplicity of times past and the quality of TV shows in that bygone era. As life would have it, I fell asleep after the episode and enjoyed a quick nap before being awakened by Gunro telling me that he was leaving the CHU area. After dinner we were told that a few of us were going to assist in a patrol that was going into Balad Ruz to bring back some detainees. The trip never materialized.

I got the word that my tattoo business had to be put on hold because another Joe had been doing tattoos on base and a few guys had gotten some kind of infection. Apparently his sterilization process wasn't all that it needed to be, or more than likely, those guys were not taking care of themselves. Either way, I was told to shut down until further notice but Big Red and Black Water 6 both went to bat for me and told the medical staff that I was good to go with my tattoo sterilization process. I only used new materials when I did a tattoo: new needles, new tubes, new ink, etc. So unless I was really careless, which I wasn't, nobody was going to get any infections from my work unless they didn't take care of things on their end. I appreciated their confidence in my work.

March 9

Our patrol rolled out at 0930, bound for the JCC so the CO could do his weekly council meeting and the First Sergeant could get some IA soldiers qualified on their weapons. As we were leaving the base, a call came from the Warrior TOC that smoke from an explosion had been sighted near CP 1. Black Water 6 gave the order for Big Red's and my vehicles to go ahead of the convoy and see if and what we could find.

As we pulled into the check point, the IA told us that there had indeed been an explosion south of their position. We passed the word to the CO and he gave us the go ahead to move south. We made our way down Route C1 and scanned the roadside for any evidence that an explosion had in fact happened. We didn't see anything for a long time. We would slow down and sign to the locals to see if they had seen or heard anything. They kept telling us something had happened further south.

We stopped a shepherd to ask him if he knew anything about the explosion and when he saw my nametape he asked if I spoke Spanish in perfect Spanish. I told him that I did and we proceeded to get information from him in Spanish. It was bizarre. We were thousands of miles away from any Spanish speaking country yet in the middle of Iraq I found a shepherd who used to take Spanish classes at the university in Baghdad. He said that he hadn't seen or heard anything. As far as he knew everything was normal.

As we were rolling along Mickey D spotted a fresh pile of rocks with wires coming out of them on the right side of the road. We radioed it in, thus beginning a five hour standoff as we waited two hours for EOD to arrive, one hour for them to inspect the sight, one hour for them to blow it, and finally, one more hour to make sure everything was clear. It was tedious. During that time we were charged with pulling security and searching the immediate area for trigger men or suspicious situations. After three hours we all had a tendency to get complacent while waiting for EOD to do its thing.

Lee Pitts, an imbedded reporter from the *Chattanooga Times,* was riding in my vehicle that day. He was a good guy, fun to have along. When we stopped to pull security around the suspected IED area, I asked him if he knew how to drive a Hummer. He said he'd never driven one but was confident that he could. I said, "Good cause if the shit hits the fan then you're gonna be driving and I'm gonna be shooting. Cool?" He was okay with that scenario. He accompanied us as we searched a few houses, taking film and giving kids candy.

More Iraqi kids

Major Ali came with the IA. I had planned to give him my K-Bar for some time but had never had the opportunity to do so. Today was that day. I had brought it with me thinking that I might run into him at the JCC or at the IA headquarters. Good fortune brought him to me and I gave my knife to him. He was very gracious although he didn't understand why I was giving it to him. Later that evening Big Red said that Major Ali was very proud to have been given the K-Bar once he had explained to him what a valued knife it was and that it was given to honor him. I hope he gets to use it someday.

The evening back at the base was low key. I did get a knock at the door and it turned out to be Terrible Teddy who I thought had already left Caldwell to head back to the States. He'd been at Cobra for his tour and was just passing through on his way home. He was the 3rd Squadron surgeon, a great guy. He came in and we talked for a bit about what he'd experienced in Iraq and how grateful he was to have been a part of it. At 2000 I took him to Chapel services.

MARCH 10

We did two patrols today. The first left at 0845. It was a trip to the JCC to provide security for the CO at the Security Council meeting that he was supposed to be at the day before but was cancelled because of our run in with the IED. The trip there was uneventful. I was the vehicle commander of A33 and we led the way with Big Red bringing up the rear.

Once we got to the JCC we all dismounted and people did various things. Two combat photographers came with us and they walked around getting pictures of what we did there. Pitts, the imbedded reporter from Chattanooga, was with us again and was taking notes at the meeting for a story that he would submit later that night. One group of guys went to work with the First Sergeant while another group gave a medical class to the Iraqi Army guys. My group assisted Handy with his construction of bookcases for the radio room at the JCC.

Of course, while we were there, we had to get bread and kabobs. We had two brothers who made runs for us. At first they were really good about it but later they started to expect tips which made me think twice before sending them. I knew it was good to give money in exchange for good work but our world had gotten to the point where everyone expected a handout for doing the minimum. Few people went above and beyond their station anymore.

After I finished helping with the bookcase and having lunch, I went to the back room and wrote a few letters. After three letters I caught a combat nap. At 1400 my driver came in and told me we were pulling out. We mounted and got in line to lead the convoy back to base. On the way back we had planned to stop in the market area and patrol, all the while looking for the guy who was selling anti-American propaganda CD's. We dismounted but had no luck in finding the guy.

It had been raining for two days at this point so the ride back was wet. At 35 mph our gunner has his body raked by rain pellets. He was prepared for the weather but it was still brutal. Black Water 6 called the convoy to a stop at the Hunter's Association building to thank them for a gift they had brought him at the Security Council meeting. When he was done, we continued our march. Word came over the radio that a suspicious vehicle had been driving around Caldwell. We went to assist the Quick Response Force, but by the time we arrived at the front gate they had already apprehended the suspects. Not needing our assistance, we headed onto base.

The route to the refueling station was covered with water again so we went mudding. It was a level route but the potholes from track vehicles were huge so it made for a good time. Speeding through them would cover the Hummer in mud and water. Z did a good job driving while I took video of our little trip. Gramps stayed in the gunner's hatch and was caked with mud when we got back to the vehicle line. I don't know why he didn't come down when we turned into the water field, but he had a good attitude about the whole thing.

We got a three hour break and then headed out for another mission. Big Red wasn't too happy about having to go right back out so he decided to make the mission shorter by driving through town once, turning around, and then coming back. From the time we met for our Op Order until the time we get back on line, we were only gone for two hours. That was nice considering we were to be heading back out at 0500.

March 11

This morning's patrol had us clearing the western boundary. We were on the line at 0450 and headed out at 0530. Priest was late so he was stuck on the 50. I played commander again. We were the second vehicle in the convoy.

After clearing the western boundary we did a 180 and headed back towards Balad Ruz. A mile down the road Big Red decided to make a right and patrol an area that we had not patrolled in some time. Nothing was going down but the rain. Iraq was like most places on this earth. When the rains came hard, nobody got out. Our patrols were a little boring because of it but I was thankful because during hard rains and bad weather we had to be so careful about driving near aqueducts. If we had been ambushed or gotten caught in a firefight, we would have had to deal with that much more.

As dawn made way for the morning and light began to break through the cloud coverage, we could see just how much of the countryside was flooded. I never expected flooding in the desert. I expected heat and sand, not cold, rain, and flooding; but that was what we'd been having since we got here. Only a few days had been good desert days and even those weren't too hot. It was a brutal environment. I did not understand why anyone would want to live here, much less die for the place.

At the front gate to the base, the drainage ditch was flooded. There was a little bridge over the drain pipes that had two feet of water gushing over it. Only a few of the Iraqi workers braved traversing the bridge in their own vehicles. We made it through with no problems. After clearing the bridge and our weapons, we headed to breakfast. After breakfast we went to the refueling station again. This time the road was completely covered by two feet of water. It was crazy. Tater drove us through the lake and only went off the road twice because we couldn't see where it was. We had just as much fun coming back through it after refueling.

March 13

There has been a lot of stress in the ranks lately. Black Water 6 thought that the only way to discipline an individual was by taking away the rights and privileges of the entire company. There have been a few bad apples in our lot and now the entire company had to pay for their actions because of the "we're a team" theory. Army leaders say they don't believe in mass punishment and yet they jockey around that by saying, "Because our unit was a team, the entire team should do the time if an individual did a crime." They used this philosophy and said it was because we should have been policing each other. This put a lot of stress on those of us who tried to do the right thing but always got lumped in with the slack asses. And the reality was this: if we attempted to discipline those soldiers ourselves, we would have been charged with assault. Why bother?

Today was a perfect example. I got a great email from my cousin. I had sent out my three month update letting everyone know how lame things were and how bad the attitudes were getting in Iraq. He sent back some great advice. He suggested that discipline should be handled in the old fashioned way – a good old, country ass whippin' if things got out of hand with a particular soldier. I agreed with him. Some of the issues we were dealing with could have been handled in house. The only problem with that was whether or not we could be certain that we would have the support of our chain of command for that kind of corrective action? If it came down to it, would they go to bat for us or would they bail on us and let us go down for taking care of something that they should have taken care of in the first place? I wasn't going to bet my career on it.

In our platoon, we had boxing matches to settle issues between guys. Sometimes they were just for fun. That was how these issues should have been handled, but most of the guys who created the problems didn't have the spine to get in the ring with anybody. This morning Boo Boo called me out. He said, "SGT Rios is a smart ass" and then continued to bad mouth me throughout our platoon disciplinary meeting. I didn't mind being called a smart ass. Hell, I was one most of the time. But when a dirt bag did it in front of the entire platoon, he pretty much called me out. What Boo Boo failed to realize was that I wasn't going to back down from any of the pukes who made the rest of our lives miserable in Iraq because of their stupidity – especially him. He didn't think I would do anything. He thought I would just sit by and let him run his mouth.

The disciplinary meeting was held as a result of everyone's complacency: failure to clean crew serve weapons after patrols, not cleaning personal weapons weeks at a time, doing drugs on duty, and so on and so forth. At our 0900 platoon meeting something was said about how a new way of doing things was going to start happening. More rights and privileges would be taken away and more discipline would be enforced. This made me laugh and cringe on the inside because that was the only path the leadership in Apache would ever take in dealing with situations. By making everyone pay for one or two individuals' selfishness rather than educating those particular individuals in one-on-one situations, they were producing soldiers who knew nothing about integrity, accountability, or responsibility.

As the new disciplines were read off, I made the comment that I felt sorry for those guys who had just reenlisted. The comment was made mainly to Pretty but Boo Boo was there also. Boo Boo kept saying, "You gotta have a back up plan. You gotta have a back up plan."

I replied, "There is no back up plan for this stupid shit."

He just kept saying it over and over like he had some great scam lined up. I responded a few more times with my comment. I guess he took it personally because his next comment was "SGT Rios is a smart ass."

I bowed my head and laughed. He said it again and then I replied, "You know, you have to be smart to be a smart ass." He shut up but you could tell that his blood was boiling.

After the meeting was over we did a police call. I walked up to him, away from everyone else, and said, "If you've got issues with me, you know where the gloves are. Just throw them on my bunk and let's go at it." Then I walked off. He didn't have a reply. One had to know Boo Boo and the rest of this platoon to know why I did that. These guys talked a lot of bull about how tough they were. I wanted to find out for myself.

God forbid I ever thought I was a bad ass but I wasn't going to back down from someone who ran his mouth. Most guys who ran their mouths were just that – guys who ran their mouths. The real guys, the men, didn't have to talk about what they could do to a person. Those were the guys I liked. They were the ones I respected. I have never respected spoiled, arrogant kids who've gotten everything they've ever wanted in life. Those guys have never been beaten down by someone with a lot of pent up aggression, bent solely on the goal of hurting them. Some of these guys needed a real good ass kicking. It would have benefited them in the long run.

About an hour later Boo Boo came back to the CHU and wanted to talk. One can tell a lot about what a person's intentions are by the first words that come out of his or her mouth. His weren't an apology. His were an excuse as to why he was in the right to say what he said. I knew right then that any apology afterwards was about as good as year old baby batter. I let him speak for awhile about why he wasn't scared to get in the ring with me, but he still wouldn't do it. He said he didn't want to settle things that way.

I told him why I wanted to fight him. I called him out because he had been stupid enough to talk trash about me in front of the platoon. Though I'm not a stickler for rank I did say that he, being an E-4, had talked badly to and about an E-5 in front of other soldiers. That was a no go and that was why I called him out. It had nothing to do with anything else. Well, okay, I did want to beat him down for selfish reasons too.

He proceeded to tell me that he was having issues at home with his wife and things were coming out in his attitude here. All I could tell him was to get over it. Everyone had to deal with issues while being away from home. His issues were nothing special and his attitude was going to get him in more trouble than his issues were worth. He finally offered an apology that I half-heartedly accepted because it had been offered so late in the game.

He brought up some issues from the past about my messing with him with a broom. When he would act stupid in the CHU I'd thwack him with the straw side of the broom, not hard; but he got the point that I wasn't impressed by his stupidity. I asked him why he didn't address those issues when they happened. Why was he throwing them out now? He had no good reply. He did say that I bullied and humiliated him when I used the broom. I'll be the first to admit that I did give him a hard time. He sometimes needed to be messed with. He was one of those guys who I thought was spoiled. He had never learned humility. The conversation ended by my telling him that I would no longer treat him any differently than any E-5 treated any E-4. I would speak to him only when the need arose and that he should do the same: speak only when spoken to.

Quote of the Day
Fog Head: "I was bowed up harder than $40 worth of jaw breakers."

<u>March 16</u>

Two more days spent at the JCC. For this rotation we had taken sixteen guys including Big Red and Dark Wing. It wasn't a bad rotation considering we stood two hour watches which were spaced out by six and eight hour breaks. I did a lot of reading. Jimmy Buffett's *A Pirate Looks at Fifty* was the book of choice. I also got to catch up on sleep. For some reason I always slept really well at the JCC.

Thus far I'd read *Where is Joe Merchant?* I was almost done with *A Pirate Looks at Fifty*. Both of those books had been very entertaining in their own rights. Joe Merchant was a fictitious rock singer gone mercenary that Jimmy had spun an exciting adventure, love story around. The story moved well. *A Pirate Looks at Fifty* was an outlet for Jimmy to share what had been going on in his life since his previous memoir, *A Pirate Looks at Forty*. He had taken a whirlwind tour south of the border and equator. His life and attitude inspired me to live life like I had nothing to lose, like everyday might be my last. Always take the road less traveled so I could write my own story.

I've enjoyed both books as I've spent the last two times at the JCC. My free time was spent in my bunk reading as much as I could and then sleeping until the next watch. Jimmy Buffett always helps me take things a little lighter. I appreciate him for that. Even when I was playing local coffee shops and Young Life clubs back home and in Springfield I would think to myself, "Maybe I could be the next Jimmy Buffett and carry on where he will eventually have to leave off." Obviously that hasn't happened yet, but every time I indulge myself in his creations my creative juices start flowing. His inspiration opens up a lot of creativity and hope in me.

I received my order today from Ranger Joe's. I had ordered two full sets of DCUs, some name tapes, three Under Armor shirts, and a balaclava to fight the sand and cold. Everything arrived in good shape and I spent some of the afternoon at the tailor shop on base getting my name tapes and patches sewn on. I dropped the tailor $50 simply because I had not given anybody any of the money I had been making with tattooing. It was a gesture of good will. It was nice being able to do things like that. I wanted to do it more often.

I sent some packages home today. The package mailing process made me really appreciate all the trouble people had gone through to send me packages. Most of the items that I received were given away simply because I had no place to put them. That was just as well. There were men and women on post who could use the items more than I could. I put my excess in a box and took it over to one of the computer centers. I would leave it on a table near the front door and by the next day most of it would be gone.

Tonight's patrol was ridiculous. Our mission left post at 2030 heading west on Route T1 and then south on Route C1. We were going to our favorite little mud hut village. We had been there many times and usually something happened. It was where we had taken fire. It was on the same route where we had found the two dead guys who had been killed by the IED.

I took point, followed by Mickey D, Little D, Pretty, and the rest of the bunch. I took us through the north side of town first; then we did some searching around the northwest side where we had never searched before. After combing that area and only finding an IP, we headed southeast through town. We passed through some alleyways and then headed south through the cemetery.

Halfway through the cemetery I moved the column west again. I had just crossed a little palm trunk bridge when I spotted three Iraqi men walking through town. I dropped to a knee and watched them for a few seconds through my NVGs. None of them appeared to be carrying weapons but we wanted to question them. When we approached, they took off running. Mickey D, Little D, and I pursued them. As we turned a corner to follow them, Pretty yelled, "Fire a warning shot!" Little D did.

At the same time, out of the corner of my eye, I saw tracers coming our way. We had just come from that immediate area where the shots had been fired. One of two things happened: either none of us did a good job patrolling the area, particularly I, because someone had fired at us from a position we had not seen, or one of our own guys fired in our direction and wouldn't confess to the shooting. Either way, we were lucky nobody got shot or killed.

Our squad came together, made a plan, fanned out, and began the process of searching houses. The first three houses turned up nothing. Of course! No one had heard or seen anything. We walked the area another time but this time we stopped in the cemetery to wait for the artillery guys to fire a couple of illumination rounds above our area. We crouched behind various tombstones, most decaying from years exposed to the Iraqi environment. The first flare went off without a hitch and lit up the entire area. The next two rounds appeared to be duds. They were fired but neither one of them popped illumination. In fact, they sounded as if they had dropped only a few meters away from our location. I was glad I hadn't been nailed by those things. That would have hurt like hell.

After the last round was fired we went back to the area where we suspected the shootings had taken place. Half of the squad stayed back at the palm trunk bridge, giving us rear security, while the front of the squad began to look for evidence of the rounds fired in our direction. I thought it was stupid. Most of the guys were just standing around conversing while looking for bullet holes in mud walls. Some were looking for expended brass, while others looked at a tunnel they believed the suspects might have dipped into. We set up security for about 30 minutes so the guys could look for a couple of bullet holes when we should have continued our patrol. What was I thinking? I was merely a medic walking point during this patrol. They were thinking about awards. All I was thinking about was that I almost got shot. What ticked me off the most about the whole situation was that it looked like the shots had been fired by one of our own guys.

After regrouping, we headed back to the vehicles. Along the way, Little D thought he should fire at a dog to thwart its efforts to fight another dog. We were finishing our patrol when out of nowhere a shot rang out. We knew it was Little D. He was the only one carrying a shotgun and he was hyped up, partially from being on patrol but mostly from being with his Broke Back buddy, Mickey D. Being hyped up was one thing but being hyped up and firing a weapon at a dog while we were on a patrol that had just "taken" fire was asinine. I wanted to smack him.

Little D got a scolding from Dark Wing. When we got back to the vehicles I told Pretty that if I ever walked point again I did not want Little D anywhere near me. This situation didn't make me feel very comfortable. Not only would I have to watch my front but now I'd have to make sure I didn't get shot in the back by my own guys.

We finally made it back to the vehicles and again Dark Wing scolded Little D on his poorly judged shot. This didn't bother Little D a bit. He was still pumped up from being able to shoot twice in one night, missing both targets. I've never seen someone so pumped up about shooting so poorly.

March 19

A cache was blown yesterday by the engineers. It was pretty cool. Atlas, Jet Ski, and I were headed to lunch when a huge blast went off southwest of our position. We looked around and saw a huge smoke cloud in front of us. We didn't know what happened at the time so we assumed that an IED had gone off. Immediately our thoughts went to our safety and then to where it was we could hide so we wouldn't have to go out and set up a cordon with the CO. Fortunately, as we found out later, it was just a cache blown by our own guys.

Jet Ski enjoying a bike ride

This morning I got to talk to my sister and dad on Yahoo! Instant Messenger. That was pretty cool. I had never used IM before. I also got to talk to Aggie. She sent some interesting pictures of herself for my own personal enjoyment. They were good. I appreciated her willingness to do that because the decent looking girls here were few and far between. None of them looked as good as she.

I also got to talk to my friend Mason Jar. She was exceptionally good about keeping in contact with me. She was 17 and one of my former Young Life kids from Livingston. Again, she impressed me with how faithful she was at keeping contact with me, more so than any of the guys whom I walked with daily while doing ministry in Livingston.

Some time between darkness and daylight someone put shit on Atlas and Jet Ski's front porch. Big Red came through the CHU area raising hell. He said that if things didn't start getting better with the relationships between the guys then he would intervene. It was about time. I went up to him later and asked him his thoughts on Code Reds – taking care of issues within our ranks on a very primal basis. Dark Wing came into the conversation so I didn't get to hear Big Red's thoughts on things. I would get his thoughts later.

March 20

Today we stayed at the JCC. It was a relatively easy stint again because of the number of people we brought with us. We stood two hour shifts and had six to eight hours of rest between watches. It was nice but had a tendency to get boring. Most of my day was spent trying to download updates for my computer but because of the phone line and power surges it was nearly impossible.

Last night we were informed that the Regiment was splitting up our Squadron into teams that would start training the Iraqi Army, so that the IA would be able to take over the military operations in Iraq. No one was looking forward to this but it was part of the process of our getting back home. I would be working with Big Red, Dark Wing, Crowfoot, Pretty, Boo Boo, Cow Tits, and Atlas. It wasn't a bad group but I could have done without Boo Boo.

I was usually a pretty laid back guy but there were certain things that annoyed me, the two biggest being arrogance and ass kissing. Most of the young guys from the Deuce were both. That was why I had such a hard time working and living with them. Boo Boo was a perfect example. His way of doing things involved both of those traits. It had gotten harder and harder for me to live with him.

Besides Boo Boo, my biggest issue was with Mickey D. When he was told that he wouldn't be with Dark Wing he immediately started whining, saying that he would be out of the service in a month. My guess was that he would very openly do drugs, sabotage something, or try to hurt himself. His attitude and reaction just proved to me how selfish he was. His whole life had been based on his selfishness and how he could manipulate things to get his way. I was saddened that such a child was allowed to wear a uniform in the first place. His intentions were not honorable and he showed no guilt about his actions. Everything he did was for self-gratification regardless of the expense to anyone else. I knew I had allowed myself to get frustrated with the entire situation.

****

From: *<Malcolm>*
To: *<Mom>*
Sent: *Monday, March 21, 2005*

Hey Mom. Everything's cool. We're on our last couple of hours at the mayor's office before we head back to Caldwell. I'm ready for a shower and some real food. We usually don't eat too much out here because of the lack of sanitation. It's pretty nasty. I took some pictures of it one time we were out here but I'll wait until I get home to show them to anyone. I want to see the faces of whoever sees those pix. They're nasty.

Glad you enjoyed John's concert. I'm chatting with a friend on Yahoo! Instant Messenger. You should try it out with Dad sometime soon. It's fun to talk like that.

****

From: *<Malcolm>*
To: *<Mom>*
Sent: *Tuesday, March 22, 2005*

I didn't know the guy who died but one of the guys from Iron Troop, the [unit] I was with, was on the convoy too and got hurt. I haven't heard anything else about the situation. Kind of sucks. That's the scariest part about going out … you never know where or when an IED is going to go off even with us looking out for them all the time.

Keep on praying for us all ….

<u>March 23</u>

JCC duty has been over for two days. The situation with attitudes finally escalated to the point that Big Red called a platoon meeting yesterday morning. We all thought he was going to smoke us for all the shit (literally – on Atlas' porch) that had recently been going on. His intention was to have an open meeting so that people could vent and fess up to anything that they needed to share. He brought along a "talk baton," which if someone had in his possession, only he could talk. We used that system a lot when I was doing Young Life.

At summer camp we had what was called Cabin Time right after Club. Clubs were set up to present the Gospel to kids. Everything done during the day was geared to give the speaker a platform from which to present the truth about God to the high school students. After Club was over, all the campers and their respective YL leaders would go back to their cabins and discuss what the speaker had talked about. Usually, everyone would sit in a circle on the floor and the leaders would ask questions about the talk. The kids would be given the opportunity to answer in any way they wanted and without the fear of anything going beyond those in the room, unless of course, it involved suicidal tendencies or molestation.

Anyway, we all congregated at the picnic table near our CHUs and Big Red began his talk. He addressed all the issues that had been raised in the last few weeks and laid out his thoughts about not wanting to ream anyone but wanting to get them help. Then he passed the baton and a few guys got some of their personal issues out. The issues ranged from Tater telling everyone that he was simply a good guy who minded his own business and wasn't mad at anyone, to my calling anyone out who had been involved in the shit on the porch scandal. Atlas got a lot of his anger out by telling everyone how ostracized he felt simply because he was older and wanted to do the right thing. He said he didn't feel like he could trust most of us. It was good to hear him finally get that off his chest. He was a good guy but the fact that he was stuck in a platoon with a squad's worth of punks was really affecting him.

The thing that really bugged me was when Mickey D got up and acted like he was the originator of the whole platoon meeting that was taking place, like he really cared about anybody there besides himself. I didn't trust him. He denied that he'd defecated on Atlas' porch. For theatrical purposes and to show that his intentions were honorable, he offered to sign a sworn statement. It was sad because most everyone there knew that he was lying and if it wasn't he who took the dump, he had something to do with it. He was definitely in on the cover up.

Later that day, Atlas came by and thanked me for what I had said, how I said it, and for standing up for what was right. I basically told everyone that if the man who did the vulgar act to Atlas wasn't man enough to admit it at the time of the meeting and I found out who did it later on, I was going to beat him to a bloody pulp. I had told Big Red what I planned to say prior to the meeting and at the meeting I told the rest of the platoon. Atlas said that most of the guys from the original third platoon were really thankful and impressed that I had the guts to say that and orient it to the guys with whom I'd come.

That afternoon I worked out with Atlas and Big Red. They smoked me. It was good. I'd been looking for a workout like that. Both of those guys were beasts in the gym but I was able to hold my own. Bong Water got in the gym for the first time today. It was good for him to be in there. He didn't work out with us but I told him what to work on and by the time his hour of working out was done, he was spent. He went again today. Good for him.

For this morning's patrol we took a bunch of school supplies to three schools. The kids went nuts when we showed up. Things were crazy enough when they saw us on the streets but when we showed up where they were with things to give them, they went ballistic! If one of our guys threw out a piece of candy, forty kids would dive for it, fighting for it like it was the last breath they were ever going to take. Of course we didn't throw things inside the school. We did that in the streets and took the paper and other school supplies in to the teachers. One of the schools we stopped at was a girls' school. They didn't act anything like the boys.

Iraqi kids

Outside the girls' school I ran into two Iraqi men who worked at the JCC. One of the gentlemen got out of his car and shook my hand and gave me a hug. It was really cool. He has been one of the kindest men I've met out here and to have him do that in public was a moment I hoped I would never forget. The soldier side of me stepped in a second later thinking, "What if this guy was just buttering up to me and then when we thought he was totally cool, he'd walk in the JCC someday with a bomb strapped to himself and blow the hell out of us?" I hated thinking that way, but unfortunately, that's the way it went out here.

Last night one of our guys had a negligent discharge. We almost paid for it today but Little D finally manned up and confessed. I was impressed that he did but he wouldn't have if Dark Wing hadn't come up to us in the chow hall and told all of us that all our MWR privileges would be revoked until someone confessed about the situation. Some time this afternoon, he told Dark Wing the truth and things were handled immediately.

I ran into Dark Wing on the way to the showers and we talked about what had transpired over the past few days. He looked spent from all the chaos. He said he was tired of covering for these guys because all they did was abuse his friendship and his leniency. Everything he tried to do to help them, they took way beyond the limits, and he was tired of having grown men act like a bunch of twelve year olds. He said that of the E-4's that he'd brought with him, his boys. I told him that I was glad to hear him finally say that and encouraged him by saying that he was a good leader. He could be great though if he would simply lead his men. The guys would follow him because he was a leader, not because they needed each other's friendship. This would be true at least for the guys who wanted to have a good leader. For the guys who wanted to abuse him and his friendship, they would whine and fall by the wayside. I hoped he appreciated what I said because he did have guys rooting for him to be the leader he could be.

I also ran into Fillnutz yesterday. He looked worn out. Apparently he had been sent on every detail outside Cobra for the past few months because of his attitude and mouth. From what I saw yesterday, he looked like he'd been humbled. He seemed to enjoy, for the most part, the missions for which he'd been slotted. He'd been shot at in Tikrit and some other places. Anytime one is a combatant and comes close to buying the farm, one reevaluates his life. I hoped things turned out well for him. I told him that he was lucky he was getting off the FOB and that if I was there, I'd want to do the same thing he was doing. We kidded that the guys at Cobra had wanted to get him killed so they volunteered him for all those missions. As we parted we shook hands and I gave him a hug. He really seemed to appreciate it; in fact, he turned around after a few steps and told me thanks.

****

From: <Malcolm>
To: <Mom>
Sent: Thursday, March 24, 2005 1:52

Now dear Mother, I've driven Hummers alongside aqueducts at high rates of speed with bullets flying all over the place – a motorcycle should be the least of your worries.

Besides, Cousin Mike has one.

And so does Becky???

March 25

We started out the day by all the E-5s and above going to breakfast together. I guess it was a good gesture for Dark Wing to let us know that he wasn't going to cover for "his boys" any longer because of all the stupid stuff that had been going on. For some reason, he still didn't utilize any of his sergeants except Crowfoot. I wondered about him sometimes; everybody did. We had the meal, and after everyone else had left, Dark Wing asked me to stay behind.

Dark Wing told me that some of the guys wanted me to quit threatening them about the whole defecation on the porch issue. I began to laugh because these guys were the same guys who ran around acting like they owned everything and everyone on post. Now they had someone who would get in their faces about issues and all they could do was whine about it instead of confronting me face to face. What little respect I had for any of them, I had just lost. I told Dark Wing that I would not say anything else but if I did find out who did it they were still fair game in my book. I also told him that I thought Mickey D was a piece of shit and had lied to everyone in the platoon about his involvement with the situation.

I also expressed my thoughts about Mickey D and Little D being behind me on patrols. I didn't trust them there. They were like two kids who'd never been disciplined and they didn't think. I didn't want to walk point and have to worry about the guys behind me, particularly those two. I don't know if that sank in much with Dark Wing or Crowfoot, but I'd finally gotten it off my chest.

Later that afternoon I began doing a recolor on a tattoo. Thirty minutes into the tat one of the guys on the CO's team came to my CHU and told me I was needed for a mission. He said that the rest of the team was grouping and that I needed to get ready immediately. That was a bummer because the tattoo wasn't finished. Bojangles said that Top had wanted me for that mission because I was a medic. I sent my canvas away, got all my gear together, and headed out to the ambulance.

Our mission was to travel to Anaconda, retrieve the body of an Iraqi Border Patrolman who'd been shot by Deacon, and then return to Caldwell that same night. The incident, from all that I had gathered, was a good shoot, a legitimate kill, and any investigation would prove that. The Deacon patrol was moving and the Border Patrolman and his car mates continually tried to enter the convoy. Their actions were suspicious. The gunner of the trail vehicle gave fair warning, following every Rule of Engagement (ROE) we were under. When the occupants of the car did not heed his numerous and escalated warnings, he lit up the car. Unfortunately, the Patrolman who got killed was the brother of one of the Iraqis who worked at our JCC in Balad Ruz.

Major Ali approached Black Water 6 and suggested that it would be a very good show of humility if we made the trip to recover the body for the family as soon as possible. That was our mission for the day. We loaded up with four gun trucks and one Hummer turned ambulance. We left Caldwell at 1500 and traveled west. I'd never been past the western boundary of our AO to that point, so for me, it was a good experience.

One of the highlights of the trip was crossing the Tigris River twice. In looks, it was just another river but as for its history, I was beside myself thinking that now I was going to be able to say I'd done that. The bridge had been blown some time in the past so we crossed on pontoons. The traffic was one way and on each end of the bridge sat an Abrams tank, ready to light up anybody stupid enough to pose a threat. The ride itself was relatively uneventful. I drove and just kept pace with the lead vehicle on the way to Anaconda. Lumberjack drove on the way back.

When we got to Anaconda, we made our way to the medical center. Once there, Black Water 6 hooked up with a Colonel and they, Bojangles, and I went to pick up the deceased while the other guys made a run to the base Burger King. We loaded up and headed to the mortuary, which was a ten minute drive across base and located on the air strip side of Anaconda. The mortuary was pretty simple. Two wood shacks housed the office and preparatory room. A semi-trailer cooler was used to store any dead personnel. After pencil whipping some paper work and ensuring that we were in fact taking the correct body, we entered the cooler and took him.

The man had died quickly, not immediately, but quickly. His eyes and mouth were still open in a look of disbelief and pain. Dead bodies didn't freak me out, I had already handled a few brutal deaths, but when I opened the body bag I thought back to the times when I knew my buddy was going to jump out from behind a corner and scare me. I knew it was about to happen, but when he did it, I still jumped. That's how it was when I opened the bag. I knew he was dead but with his eyes and mouth open he looked like he was crying out for someone to get him out of the bag. Then my mind went wild reflecting on all the old horror movies I'd seen in the past. What if the guy got up in the middle of the return trip and banged on the cab door of our vehicle? What if he sat up while we were moving him from the cooler to the ambulance? We would have shit a brick. My mind was such a terrible thing to waste.

Our return trip happened fast. Having no traffic, we were able to move quickly back to the Balad Ruz JCC. When we got there the mayor was waiting for us along with a large contention of ING, IP, and friends of the deceased. It was very sad for everyone. The mayor wanted us to follow him to the hospital so that we could leave the body with the family. We followed a ten car procession of ING and IP. It was a moving gesture of love for this guy and his family. We unloaded the body to the sobs and cries of a few of the men; and after the CO wished the mayor the best, we headed back to Caldwell, everyone still in a very sober state of mind.

March 28

We got back from two days at the JCC. We spent Easter there, numbed to its religious significance by a barrage of movies that helped us pass the time. The greatest thing about Easter this year was the sunset. I went on watch at 1900, just as the sun was setting. Sunset continued to be a magical time of the day for me. To me, it was when God truly painted the universe with His glory. Not only did I get a show of beauty to the west but on nights like this one, I was able to get an encore in the east.

As the sun set I looked for the moon which I knew would be near full. It didn't rise for some time. Then, from out of the east, over the mountains that were the natural border between Iraq and Iran, came a huge, flaming orange ball. It was awesome! It looked like the Great Pumpkin that Linus always told Charlie Brown about. If that's what Linus thought the Great Pumpkin looked like, Charlie Brown should have stayed in the pumpkin patch. It was worth it. I kept laughing about that thought as I watched the moon rise in all its glory. For the first 20 or 30 minutes of the show it was burnt orange. I just kept thinking of all the things I was thankful for and how that was such a good way to end Easter, a clear reminder of the awesome power of the resurrection. After 30 minutes the moon began to change to its regular grayish, white color. It was still larger than life and stood out spectacularly against the backdrop of stars which seemed to be its audience.

During that time I also thought about being home with my friends on a night like this. It was warm and the moon was huge. We would have been at Sunset Rock, rappelling down the 80 foot face, smoking cigars, and sharing dreams and jokes with people we really enjoyed being around. How I wished I could be back there. "Soon enough," I kept telling myself, "soon enough."

This morning we cleared the western boundary. On the way back to the JCC we stopped at a local vendor and got some eggs for Crazy. While we were dismounted, I noticed two guys skinning a sheep. I ventured over and took some pictures. They and their friends had a good laugh at my expense. I suspected that they thought it was hilarious that I, an American soldier, thought what they were doing was worth getting out of my vehicle for and taking some pictures. It was a good break in our routine.

Quote of the Day:
Dark Wing: "He's walking around lost like a blind fag at a wiener roast."

****

From: <Malcolm>
To: <Mom>
Sent: Monday, March 28, 2005

Hey Mom,

Easter's over now. Not too much went on. We were at the mayor's office standing guard. It went pretty slow. Watching the moon rise last night was the highlight of the two days for me. It was awesome! Huge. Orange. It looked like the great pumpkin. Made me wish I was at home getting ready to go night rappelling with Jack, Joey, and all the LA boys. Oh well. Dreams are still good.

MARCH 29

Today I rose at 0445 for a 0530 muster at the Bradley line. Our mission was to clear the western boundary and then do a foot patrol in an area southwest of Balad Ruz. We ate breakfast before we headed out. After clearing the boundary we pulled off the road to fire crew-serve and personal weapons. We had a good time unloading a few rounds into cans and water bottles. The Bradleys shot at a nearby dirt mound with their big guns. The noise was deafening yet very exhilarating. The sound and feel of what it would be like in a fire fight with a Bradley only 15 feet away was crazy. No wonder guys came back with hearing loss and I had earplugs.

After we cleared a few magazines, we mounted back up and headed to our foot patrol location. It ended up being down one of the aqueducts that runs south of Balad Ruz. Gunro was convoy commander on this excursion. It was a stupid mission. We patrolled down the aqueduct in the open, just plugging along with Pretty walking point and me second. We got bored with the patrol rather quickly but then we spotted some ducks in an adjacent field. Pretty took aim and shot twice. I shot once. Fortunately for the ducks, it's harder to hit a duck with 5.56 rounds than it is with buck shot. Neither of us hit but we scared them. Pretty was a big duck hunter, so naturally we ragged him about missing the remainder of the day.

After lunch a few of the guys decided to lie out on top of their CHUs. Boo Boo and Little D were catching rays along with one of Little D's new female friends. She was rather large and quite ugly, as were most of the women he had relationships with, so I thought it would be funny to get a picture of them lying out together. I got Boo Boo to take the picture. I went back in the CHU and got ready to play volleyball with some of the other medics.

I went to CFB and Lumberjack's CHU to see if they were ready to play. CFB wasn't there so I lay down on his rack and caught some sleep until he showed up about an hour later. He and Lumberjack got dressed and then we headed to the volleyball court. We played for a few hours. A team of Iraqi Army guys came out about an hour into our match and we began playing them. They beat us twice.

After the game I hit the showers and then went to the mail room and picked up some mail that Big Red said we had there. I brought it back to the CHU area and noticed that after getting their mail, Boo Boo, Dark Wing, Mickey D, and Little D quickly got away from me. I didn't know what was up but I had a feeling that something was. It wasn't unusual for these guys to shun somebody who didn't play their game. After they left Jet Ski called me over and said that he had overheard the guys talking about moving Boo Boo out of the CHU and putting Gunro in there because he lived by himself, away from everybody else. Apparently it was because of some pictures the First Sergeant had seen and I was being blamed for them being shown.

I said, "What the hell?" and went back to my CHU to change for dinner. Most of my dinners of late were done alone since I wasn't "one of the boys." That was fine by me. I had walked alone most of my life. I was fortunate to have some great friends whom I knew I could call on at any time and they'd drop what they were doing and be there for me. Those were guys I appreciated. Those were guys I could count on. These other guys were a joke.

Chow was spent sitting alone, watching other soldiers, wondering if they had to deal with this kind of crap. Crowfoot, Pretty, and Big Red came in and sat at another table. No contact was made between us until I got up to leave and almost ran into Big Red. I greeted him and then headed out. Unfortunately, he was starting to allow the attitude of the Deuce to rub off on him. He had the ability and the drive to be a great leader but with their influence he lowered his standards of excellence. I was beginning to lose respect for him. I hoped and prayed that I didn't lose it all.

I passed Priest on the way back to my CHU and told him it was cool for him to come over and get the tattoo he'd been asking me for. He got his wife's name in Arabic tattooed on the back of his neck. It was a pretty simple affair. While I was setting up he told me in confidence that the rumor was out that I was the one who got Dark Wing in trouble for his allowing Little D's girlfriend on the roof of his CHU to sunbathe. Apparently I had shown the First Sergeant some pictures of everyone up there with little to no clothing on. Of course I got the blame because I had gotten Boo Boo to take one picture while I was around.

This was the point where I got really frustrated with how things were going with our squad. For starters, I didn't do anything wrong. Secondly, all the guys were blaming me, from what I was told, yet none of them had the courage to ask me about the situation point blank. Their big game was to start talking behind one's back and get everyone else in the squad fueled up so that regardless of the truth, their little group was mad. The saddest part of that was that there were a few guys as old as I or older who rolled that way as well. Most of the guys in the group were all under 25 but they acted like they were still in high school. The ridiculous thing about that was that out of all the high school kids I knew, most were more educated and more mature than these guys.

Before I started Priest's tattoo, I confronted Little D about the situation and he flatly denied knowing anything about it. Boo Boo was there and didn't have the balls to chime in either. Next, I sought Dark Wing. Oh, wouldn't that be fun, confronting the Platoon Sergeant. It actually wasn't that bad. By the time I found him, he was in Little D's CHU and he was fine with talking to me about the situation. I told him that I had taken a picture but had not passed it on to anyone. He said he didn't think it was me now because Top had not mentioned my name. I also showed him the picture that I had taken to see if it was the same one that he'd seen. He said it was a different picture and from a different camera. He also told me that he was going to move Boo Boo out of the CHU and Gunro, in fact, was going to be moving in.

March 31

It was the last day of March. It was hard to believe that we'd been in Iraq this long. It was even harder to fathom that we were going to be here much longer. Yesterday was the most laid-back day I've had in a long while. Most of my day was spent reading Dan Brown's *Angels and Demons*. The things that intrigued me the most other than the great story were the ambigrams that were throughout the pages of the story. Every time I saw one I thought it would be cool as a tattoo for my arm or back. I'd love to get five of the six ambigrams that were in his book. The ambigrams for Earth, Wind, Fire, and Water were awesome. The sixth ambigram in the story had the four elements of science as one diamond shaped ambigram. It was the best one.

Yesterday Boo Boo moved out of the CHU. I didn't offer any help and he didn't ask. He moved rather quickly. I was impressed. I was happy that he was out. The atmosphere immediately got better.

I was the black sheep of our squad. I still talked to the guys but no one hung out with me. That was fine. I'd been in situations like that before, twice: once when I quit drinking and the other time when I became a follower of Christ. It was still funny and sad to me how people would run to you when you were doing what they thought you should be doing and when you didn't, they'd run like hell. Standing up with Atlas that day put me at odds with all the guys from the Deuce; and even though Atlas said the other guys in the platoon were inspired by what I said, they still did not hang with me unless they needed a tattoo. I was happy that I was bold enough to stand up for what I believed to be the right thing. I prayed that I'd be man enough to do it again.

When I was young, I went to a summer camp in Monterey, Tennessee called Camp Country Lad. Malcolm Williams, the father of the man who ran the camp, was the person I was named after. Malcolm II was and is a great man. I've never questioned his heart. He has always served God unashamedly and followed his heart with his ministry to boys. His camp was and is a getaway from the onslaught of modern society. All the cabins are restored log cabins from over a hundred years past. Most are closer to 200 hundred years now. He brought boys back to the basics of camping and living. Nothing fancy at the camp but the spirit of God roaming the trails, the lake, and the adventures.

One thing Malcolm used to say was, "Have fun in the right kind of way." I used to love hearing him say that because I could see in his eyes and tell in his voice that he meant it. I never heard him get down on anybody directly. He would always question the person in a manner that made that person take ownership of his errors. In doing that the person became accountable for his actions. He owned up to the responsibility that came with correcting that error. Malcolm's method also ensured that the person learned from both the mistake and the corrective action. He was the first person whom I could remember challenging us to reflect on the choices we made in life. That was one of his greatest traits – the ability to teach through loving discipline.

The way he carried himself often came back to my mind when I found myself in situations that I should not have been in and even in those where I should. How should I carry myself? I believed I was acting in the right kind of way. Why shouldn't I be bold enough to stand and deliver? It was easier to do the right thing first than do the wrong thing and have to go back and try to fix what one had done wrong, wasn't it? The beauty of his philosophy was that he was human and he knew we all were too. There was a lot of room for mistakes and major foul-ups, yet as with God, his grace was abounding. I've done some bad, stupid things in my life but Malcolm was one of those guys who always seemed to be there to encourage me back to the right track. When I finally did get back on track, he continued to be one of my biggest fans. I wished he was with me so I could have an example of grace under pressure.

You Can Run Your Mouth but Can You Run the Ball?

April 3

The past few days have been a mixture of whirlwind missions and killing time on post. Our first mission was one that we had trained to execute with ODA. We were to fly into an area south of Caldwell that was a hot bed for insurgent activity, secure it, and allow ODA to flush out bad guys, either as detainees or by killing them.

It had been a few weeks since that particular mission was supposed to go down. It had been postponed because of weather, sand storms, and to some degree, the lack of presence of the high tier insurgents whom we were supposed to snag. When the mission finally did go down, our CO gave the air assault part of the mission to another platoon. The mission became a night op. All the guys in our platoon were upset that we didn't get to do the air assault. All we did for the operation was convoy to the location of the insertion in our Hummers and provide an outer cordon to block off anyone's entry to or exit from the immediate area.

The only thing cool about the mission was that we got to see the Black Hawks and Apaches doing their mission in black out. It was an amazing thing to know that American pilots could fly those birds with night vision equipment as effectively as if they were in the best conditions in daylight hours. I kept thinking to myself that if I had been an insurgent, I would have run like hell the second I heard the birds. They were looming black phantoms! It was eerie. I was thankful they were on our side!

As we were loading up for the mission we heard artillery leaving our base. After nosing around for a bit we were able to find out that there was a fire fight going on between insurgents and the Iraqi Police in Balad Ruz. Our team was anxious to go but our leadership was focused on the mission in the south. Our route to the south took us past Balad Ruz. We were all chomping at the bit to go into town and get in on the action. Our convoy continued south. During the firefight the insurgents were able to kill a Colonel and Major of the IP.

First platoon did the assault with ODA. Some Iraqi Army soldiers rode along as well. By all accounts, the mission was a success and we came back with a few detainees. These particular insurgents were high profile targets in our area of operation so for us, the mission turned out to be a mission worth doing. Besides getting a few bad guys off the streets, we were able to disrupt what chain of command the insurgents did have. Sometimes on these missions, we would be out for hours and we would not find anything or anyone. We would return to base ticked off because we felt like we had wasted the day. It was always good to bring home a "trophy."

Last night another small gunfight took place. This one was closer to the JCC in Balad Ruz. Some shots were fired. Most of our day was spent on base. We had been scheduled to do an operation in Balad Ruz to flush out a vehicle-borne IED maker but it was given to another team. They ended up having a very successful mission. Not only did they get the target they were after, but they also snagged 13 other people who were involved in the bomb making process. I wanted to be on that mission. Whether it had been a success or not, it was always good to be off post to get away from the BS that went down.

Instead of being a part of that mission I sat around the CHU playing my guitar. I wrote a few songs. One was a blues song in the style of Stevie Ray Vaughn but it didn't have all that amazing guitar work of which SRV was famous. I'm just a rhythm guitarist. It was still a good song. Maybe one of these days, after this whole thing was over, I would get it down to Sinsation Studio to record.

Today we finally went on mission again. Four Hummers loaded with 20 soldiers headed to the JCC in Balad Ruz to meet with some of our Iraqi counterparts. Our mission in Iraq had changed to that of overseer / advisor for the Iraqi Army. The hardest part of doing this mission was that most of the IA had seen more combat than any of us. Some had fought in the Iraq / Iran War, the Gulf War, and now, Operation Iraqi Freedom. A large number of them had transitioned from Saddam's Army into the New Iraqi Army. They had seen war.

They were good men, willing to fight and die for the freedom of the people of Iraq. They, like us, were doing a job, mostly to support their families and to survive. When we had the good fortune to relax and talk simply as men, we were able to laugh and see each other as that: simple men who just wanted the whole thing to be over so we could finish our lives and be with our families and friends.

One of the Iraqi Army soldiers who also acted as an interpreter, Abas, had become a good friend. He had been studying to become a mechanic prior to joining the New Iraqi Army. And why did he join the NIA? He said that he joined to have a job but now was afraid to get out because it was the only place he could make money. If he got out insurgents might come looking for him because he had been in the NIA helping coalition forces. His chain of command might come after him because they would say that he owed the NIA more time. Either way, it was safer for him and his family to remain a soldier. The funny thing about Abas was that he seldom carried a gun. He wasn't much of a fighter. He was a peacemaker.

Abas did an incredible job as an interpreter for his unit and we enjoyed having him around because he gave us a way to communicate. We wouldn't have been able to had he not been there. He and some of the other soldiers invited me to the IA station often for lunch. We would sit and talk. We always ended up trading things like gloves, flashlights, and portable DVD players for knives, uniform items, and occasionally, a bottle of liquor.

APRIL 4

The day started out pretty much like any other we've had here in Iraq; not much was going on in the morning but a patrol was slotted for the afternoon. I spent some of the morning finishing a movie that I had started the night before and then went to the Internet room to check emails, IM, etc. At some point before lunch I tried to get OB to slide me one of the new medic leg pouches. He was able to square me away and I spent some of the morning restocking my medic bags.

When I first got in country I took my regular aid bag with me on every patrol. Obviously, I never knew when something might go down so as a medic, I always wanted to have enough meds and aid on hand should something happen. But after a few months of lugging all that extra weight around, I got smart. I'd take my big bag with me but I'd leave it in the Hummer for major emergencies and would carry either a small butt pack or a leg pouch with just the basics for a couple of trauma treatments. It turned out to be one of the better decisions that not only I made but a few more of us that were going on daily missions made. The smaller, lighter bag made us more mobile.

Later in the afternoon I finished repacking my original aid bag and then headed down to the motor pool for our Op Order and SP. On this mission I would actually be utilized as a medic. Lima Bean was our patrol leader. He briefed us on the mission and then the platoon sergeant, Jonesie, led us in a reading of Psalm 91. Finished with orders and prayers, we loaded up in three Hummers. We ran east on Route T1 to a town called Jizr Naft where we took a left and headed north to our destination, White Castle.

White Castle was an Iraqi Army outpost above a checkpoint that they controlled as well. It was a two story white building. The IA stood watch at the building and at the checkpoint. There wasn't anything spectacular about it, just a building in the middle of a few huge, sprawling fields.

Far be it for me to pass up a good opportunity when it became available. I traded some dollars for some denars and got two knives as well. I had been collecting the IA knives so that I could give them as presents to friends back home. That was the highlight of the mission while we were there. After an hour or so of meetings and a massive assault of gnats on everyone standing over watch in the Hummers, we remounted and headed back to Caldwell.

When we were four or five clicks outside the gate to Caldwell, a call came over the radio from ODA saying that they and a force protection team, as well as a unit of IA, were pinned down in a fire fight. They had been ambushed and were taking automatic fire and RPGs. Immediately we asked Lima Bean if we could respond. Everyone in the vehicle felt like we should do something to help our brothers. Lima Bean got off the radio and told us that a quick reaction team was already heading out the gate to assist. As we closed in on the gate, we saw the other convoy leaving. We pleaded again with LB to allow us to link up with the reinforcement team. He said his hands were tied. None of us knew what the next twenty-four hours would be like.

We pulled into the motor pool after refueling and I was dropped off as the driver needed to do some maintenance on the vehicle. My mind was racing as I tried to grasp what was going on down south. We had no information. All we knew was that our guys were being shot at. As I walked back to the 1st Squadron aid station to see what they had heard, I just kept thinking about our guys being pinned down and what they must be going through. Did they have enough ammo? Were they able to find cover and still be able to mount an assault? Could help arrive quickly enough? Was anybody injured? What, in fact, was happening down south? What would I be doing if I were down there? Would I do my job? Would I freeze? Would I be the hero of the day?

No sooner than I started talking to Cankles than Top ran by me and asked if I was the medic going with him. I said I had just gotten back from a mission and wasn't slotted to go but I'd be more than willing to go if he needed me. He told me to follow him and we'd see if his medic showed. Much to my delight, no other medic showed up so Top made a seat for me in his Hummer. He was organizing the second support team.

Because of the breakdown in our platoons due to the new mission our unit had, it took quite a bit of time to get enough Joes together to fill the needed slots on the convoy. Three Hummers, a Mortar 113, two Bradleys, and three tanks rolled out together. Our blood was racing as we were all eager to get to the action and help out whoever was down south now. But just like every other time we'd assembled for quick runs, Mr. Murphy reared his ugly head. As we got on Route T1 and headed west, we could go no faster than 15 mile per hour due to the Mortar 113. It was killing me! Everyone in the convoy was going nuts. How could we assist our team mates if we couldn't even get to them on time?

As we cleared TCP 4 we heard reports of other support teams who were getting ready to join us all down south. The post was alive with men willing to go out and help brothers in harm's way. We cleared TCP 1 and headed south on Route C1. It was on Route C1 that we found out how bad our situation really was. Not only were we going too slow to assist the ODA, force protection, and IA soldiers who were in the firefight, we got passed by two convoys that had left after us, and we heard that four Americans were wounded. Knowing that made the crawl of the convoy that much more unbearable. It was like pulling teeth without anesthesia!

Our convoy made its way to CP 12 where we set up a defensive perimeter and waited for the retrans team, the Paladin teams, and the refuelers. While there we heard more news about the continuing battle. Some of the IA had gotten hit. Dust-off birds were called. They were coming in with the confidence with which only those pilots could fly into a hot spot. All we could do, as the heavy ground reinforcement, was listen to the radio transmissions. Chatter covered the airwaves like an epidemic. It seemed like everyone in squadron was on that radio frequency.

Another hour passed. Word got to us that another convoy which included the Paladins was on its way. The Paladins were artillery capable. They would be used to illuminate a given area from miles away or lay waste to it if need be. They finally made it to our position and rallied around our already established defensive perimeter. Top was coordinating all the functions within our rally point.

At 2200 another support team was loading up and getting ready to head our way, this one with more medics and my actual combat team. It was at that time that we were told to move to another position south of us so that the Paladins could be within range to utilize their arsenals. From the radio transmissions we could hear that dust offs were continuing to be flown from the fire fight as the last remnants of the AIF were being neutralized.

We set up our next battle position in the same area where we had done the "Tomb Raider" mission in January. We were in an open field and set up in a semicircle so that the Paladins could fire south of us. There was absolutely no cover but we had superior fire power should anyone try to come at us from an open field. After securing the position, our gunners began to pull over-watch, scanning the surrounding area with NVGs. Top continued trying to convince our squadron commander to let our team move closer to the battle but SCO said that we were not needed; they had everything under control. Besides that, he said that he had superior air support and did not want illumination rounds fired over his position. The rounds would silhouette his vehicles and men as well as get in the way of the air support. We sat in the field for the rest of the night listening to all the action that was going on south of us.

The whole time I was on this "reinforcement mission" I was thinking about the medics and men on the ground. Who was down there? I knew that the ODA Doc was there but I could only think of one other medic who might be down there because he was on a force protection team, Georgia. As far as I knew, those two were the only active medics down in the fire fight. Fortunately we knew from the radio transmissions that wounded were being treated and flown out and those same birds were dropping off fresh medical supplies, ammo, and water for our guys.

From what we had heard up to that point, everyone did his job exceptionally, thanks in great part to the professionalism of ODA, as well as the calm demeanor of the majority of men involved in the fire fight. The terrain where the men were fighting was a series of ravines and berms, and soldiers were on one side of a berm or aqueduct while an AIF would be directly across from them behind a bush or pile of dirt. The waterways were deep enough that the Hummers had trouble advancing in some places. At one point, one of our soldiers was hit. In the line of fire the CO of the ODA team ran out and retrieved him. One of our company officers from Apache, Rugger, took a round from an AK-47 to his left shoulder. He was an old rugby player and the stories that came back later were that as they were taking him out of the action he was trying to get pictures of his wound because he said, "I've been hit harder than this."

Rugger giving the thumbs up after getting shot

When Warrior Six made the decision to call in the air strike, he was given Apaches and an AC-130 Specter Gunship. I wish we could have made it all the way down to see the power those birds really had. It would have been awesome! They were the clean up crew. Warrior Six wanted to finish this thing. After ensuring that everyone was pulled back to a safe distance the air assault began. We could feel the concussions of the impact and explosions at our position 15 clicks away. I could only imagine what it would have been like to be right there.

Later, in talking to some of the guys who were there, we found out just how low they had gotten on ammo and water, two things that were a necessity in a sustained fire fight. Fortunately they had just about wrapped things up as they reached the last of their ammo. The last kill that we heard was an insurgent who was holding out with the fleeting hope that he could stop the American advance or maybe secure himself one more virgin if he went out in a blaze of glory for his cause. The air strike was called in on him and he got the opportunity to experience hell on earth.

One of the few funny moments of the night came after the bombs lit up the last insurgent. Warrior Six asked ODA if they thought he should send some Joes over to the area to make sure the guy was dead. The ODA radio man, bewildered, replied, "Sir, I don't think you're going to find anything out there but pieces of him." All the rest of us had a good laugh at the SCO.

Another funny event that we heard on the radio was the spotting of a four legged creature walking around the immediate area where the firefight had taken place. Someone had spotted the beast using his thermal imaging equipment. Warrior Six asked what it was they were looking at and the responding station said, "We think it's a dog, sir." Warrior Six then shouted back frantically, "Shoot the fucking dog!" They did and that was the last of the gunfire for the evening. In the morning light the teams found out that it was not a dog at all. Our guys had smoked a camel.

The night finally gave way to daylight and the area where the fire fight had gone down was secured. AIF bodies were found and counted and weapons were identified, gathered, and then taken from the area. Overall, the mission had been successful. As for the human toll, four Americans had been flown out wounded in action, approximately eleven Iraqi Army guys were airlifted out, and approximately fifteen insurgents were killed. One camel bought the farm.

After the area was cleared by the assault team the word was given for us to rally and then head back to Caldwell. Top lined us up and we convoyed back to the refueling site. Once there we dropped off the Paladins and Cats so we could move a little faster. Besides, they had enough fire power to take care of themselves this close to base. We escorted a liaison officer and his crew to a meeting in Balad Ruz with some sheiks prior to making our way back to post. After dropping him off, we quickly headed back to Caldwell to find out what we could about our soldiers.

When we got back to base, we heard the bad news. Two of our brothers in arms had died of wounds on the battlefield. I hadn't known either one of them but felt sad at the loss of two American heroes. Both were family men and both had established deep relationships with the men they worked directly with during their time in Iraq and the National Guard. These were the first deaths from FOB Caldwell. Another 278th soldier from another FOB had been killed by an IED two weeks previously but this was close to home. We were still concerned about the welfare of our wounded LT from Apache.

Looking back over the last 24 hours, we were very frustrated with our part of the mission. First, I was a medic who was not utilized on the battle front and could have been there at anytime. Second, it seemed like only officers were in on the action. Granted, all the officers had crews of enlisted with them but those of us who had been training and patrolling the area for the past four months were not allowed in the area where the action went down. I thought it was a sad statement about our chain of command.

From a ground-pounder's perspective, all they wanted to have was the glory of having been in a combat situation. It seemed, again from our perspectives, that all they were seeking were the accolades that would come from their participation in a firefight or in the aftermath thereof. This was the big show for these guys. The more ribbons and awards they could come back with – that they had put themselves in for – the more opportunities they would have with promotions and positions when they returned home.

Third, none of the assets that we had so painstakingly gotten together for the assault were used: the Bradleys, tanks, Paladins, and mortars. Not only were they not used but because of having to wait for them, we were not in a position truly to assist our brothers in that hostile environment. Even Big Red had something to say about the operation along those lines. Fourth, the soldiers who were supposed to be working our company radio room had all dropped their jobs to go out on the mission. They did not need to. We had enough personnel to do the job. But because of their working relationship with officers and senior enlisted they were given the opportunity to "earn a close action badge." But their job was in the TOC, not out on the mission, and they failed at their job.

I guess we all had our gripes and complaints about the whole ordeal but it came down to one thing: two American brothers had given their lives in the procurement of freedom for the Iraqi people. My heart went out to their families and friends. I could not imagine what it would be like to lose a loved one in combat. Even being a single 35 year old, I am sure it would be devastating for my dad, mom, sister, and all those who I call friend.

Tribute to Steve Kennedy

****

From: <Malcolm>
To: <Mom>
Sent: Wednesday, April 06, 2005

The rumor is half true. Two 278th guys were killed and two wounded in a fire fight two days ago. We were on our way there as it was going down. It's sad here. One [soldier] in particular was very well liked and known by a lot of guys on base. There were a lot of heroes down there that day. Don't worry about me. Just pray for their families and for the two guys who got wounded.

April 8

I've been busy with patrols, meetings, and now, medic recertification training. As far as patrols go, I've been on eight in the last five days. It's been crazy but at least it keeps me busy and out of the CHU. My new roommate hasn't been any better than Boo Boo. Normally I'm pretty easy to get along with because I try to mind my own business. I don't like people butting into mine.

Gunro was not necessarily a bad guy. For the most part, when he was one on one he was decent. He had happily helped me out a time or two. The issue was when he was in the midst of a group of people that he became the proverbial pain. Gunro was that former Marine who never got to be a career Marine. Word around the horn was that he only got to do one tour and only made sergeant. Then for some reason he got out. He has lived the rest of his life being mad to some degree at whoever did not allow him to retire as a Marine. All he did and thought in Iraq was USMC.

I had no problem with Marines. In fact, all of the Jarheads I had known over the years had been great guys. They were all proud to have been called Marines but they were also humble with their pride. Gunro was the exception. He wanted to be called Gunny but he never was a Gunnery Sergeant much less a Staff Sergeant in the Corps. Every decoration he had in the CHU had something to do with the Marines. He sat in the CHU watching movies and wearing his DCU pants, brown shirt, and a USMC patrol cap. Sometimes I just wanted to jump up and yell at him, "It's over! Move on man! Get out of the past. It's dead. You were a Marine but now you're in the Guard!"

My biggest struggle with him was that he was bitter about life. Very seldom would he smile. Very seldom would he go out of his way to have uplifting, energizing conversations with other soldiers. He had pissed off everyone in our platoon so we added fuel to his fire any time we could. He didn't even smile at the other soldiers on post. It seemed to me that he felt like the whole world owed him something. Anything that did not go exactly as he wanted it to go he would bitch and moan about even though it might have benefited him.

It was sad but anyone could read him and know it wasn't just because he was in Iraq. Something in his bitterness burned deeper and longer than any stay in Iraq could have. Nobody wanted him as a leader and when a leadership position opened up, the spot was given to someone else. He became even more bitter and critical.

Patrols and missions have been pretty normal. Yesterday I did help teach a first aid class to ten Iraqi Army soldiers at the JCC in Balad Ruz. It was a good time. Eddie and Mr. T did most of the teaching. I worked on a Power Point presentation that could be used later when we taught. I taught the segment on Casualty Collection Points. When I talked to the Docs about their experience in the firefight on Monday, they told me that one of the biggest problems they had experienced was the CCP. It was mainly because the IA did not understand the concept of bringing their wounded to a centralized location for security, evaluation, treatment, and evacuation. In any mass casualty situation a central, secure area must be in place for the wounded to be treated. I say secure, but any one who has been in combat knows that secure may not always be the case.

Today's patrol was with Lima Bean. We rolled east to White Castle. I ran into some of the IA that I had traded with before. I bought two more knives, one of which was a 1913 Remington bayonet. I got quite a few compliments on my purchase and even had a few offers to purchase it from me. After leaving White Castle, we headed northeast to IED Alley.

The area had hills. These were the first hills I had seen since leaving the States. I missed the hills of Tennessee, and being in this part of our AO brought that part of home back to me. While everyone else was talking about where to set up observation points for sniper teams, I was standing alone on a hilltop breathing in the air and dreaming about hiking in Big South Fork, Standing Stone, and Camp Country Lad. It was awesome.

After scanning the area we headed west. Our next mission was to recon Balad Ruz as Lima Bean had yet to conduct operations there. It was a smooth recon. We entered the city through the market area, drove west through town on the main road, and then turned south on the street where the mayor lived. After driving through more of the side streets we came out on the backside of the market area, turned east, and then headed back to Caldwell.

After chow I returned to the CHU to rest for a bit because today's patrol was the first day patrol on which I had almost fallen asleep. Usually I drove but this patrol I just sat in the back and didn't talk much. I knew I was tired from all the missions and constant movement so I needed to let my body and mind rest. As fate would have it, Gunro came in raising all sorts of hell and kept me awake the rest of the afternoon.

I went to dinner and afterwards attended the second of 52 medical classes I had to do to get recertified. It wasn't bad. A good looking medic was teaching the CPR class. I had run into her one night while we were in Shelby. She was fun but for some reason we had never really hung out. She had beautiful blue eyes.

April 14

The last few days were busy until yesterday. We were at FOB Knot working on the fire range. We didn't actually do any of the work. We provided security. Handy and Cow Tits worked with a squad of IA to finish up some work they had started a few days earlier.

We left Caldwell at 0730 and got caught behind another convoy, which in turn was slowed by a wreck a mile from the front gate. An IA truck had rolled with a few passengers. None of us knew the extent of the injuries involved in that wreck but the vehicle looked pretty bad.

After clearing the wreck, we continued on our road march east to the FOB. We were there for about 20 minutes when we decided to try out the new targets. We fired off a few magazines of 5.56 from standing positions. When we were finished we walked to the targets to see the effectiveness of our shooting and the condition of the target boards. Everything was golden. The targets on the boards were the normal, paper silhouettes. The targets on the ground were made of "hillbilly up-armor" plating. Handy had acquired some and had cut out silhouettes for us to shoot.

A few of the IA soldiers came up after we were back at our shooting position and asked if they could shoot our weapons. I cleared it with Dark Wing and watched as four of our Iraqi friends cleared two more magazines. They were not the best shots in the world but they were able to get some fire time in with American weapons. We decided to stop shooting when some of the bullets ricocheted over our heads. Bullets on steel, at an angle, were not a good thing.

IA on the firing line

For the rest of the time we were there, Dark Wing, Crowfoot, Pretty, and I sat in our Hummer and told stories of our pasts. It was good and helped to make the time move a little faster. Abas joined us and told stories of what Iraq was like under Saddam's regime. He said it was horrible, with tens of thousands of people murdered, families removed from their homes and relocated to new towns, and soldiers kept in service by threat of jail time.

He spoke of the reasoning behind the IA's use of violence for interrogation. The Iraqi people understood violence. To get information that they wanted, the IA would use violence or the threat of violence first and then ask questions. All of them said it made the acquisition of information much easier. That was something the Iraqi people understood. They didn't volunteer information. They were tightlipped about everything.

Abas also told about the "law of reciprocation" in Iraq. He said that if he were to kill another man and no witness came forth, nothing would happen to him. But if someone found out about the killing or witnessed it then the family and friends of the victim would come to his house and require a payment for the life. The offender would have to pay the debt with money, women, or the giving of his own life. Ultimately, it all came down to killing: an eye for an eye.

We finished at the fire range, loaded up, and headed back to Caldwell. We were able to make lunch, and after lunch I left the rest of the guys and attended a medic recertification class. As a field medic I didn't always get to do the things the medics in the aid stations were able to do, so it was a good refresher for me.

When the class was over I ran into Fillnutz. He was in from Cobra on his way to Germany to have surgery on his knee. He ended up hanging out in my CHU the rest of the day. It was good catching up with him. In the past when I'd hung out with him, there was friction because we were so similar: headstrong, sarcastic, and proud. Since coming to Iraq I had seen his heart change somewhat as he found out what humility meant.

The past few days were filled with other missions to Balad Ruz and FOB Knot. I usually liked to go to the Balad Ruz IA station because I got to see some of my new friends, but after awhile it had been a bit much because every time I was there the guys were asking me to get them something: DVD players, cameras, etc. I didn't mind doing it for a few of the guys, but when everyone asked for something it got hard to deliver.

Usually when I was at the IA HQ I spent time with LT Ali, LT Ahamed, LT Saddam, and Abas. They were all very entertaining and made me feel like they appreciated my being there. They always had a good story to share about life in Iraq. LT Ali always wanted to give me something. He offered gold, whiskey, porno, and even his cell phone. He was the wild one of the bunch - a good guy but wilder than a three peckered goat. I liked him from the instant I met him.

Gunro's attitude finally got the best of him today. He tried too hard to undermine Bong Water Blue's position as platoon sergeant in their team and was booted by the First Sergeant to the JCC squads. For now, he will be working five days in Balad Ruz and three days in Mandalai. I hoped that he would see the error of his ways before he did something else and get himself even deeper into the pool of shit in which he was already swimming.

April 15

Tax day in the US. It's been a slow day at Caldwell. Some of our guys were slotted for a big mission spearheaded by ODA, but because of the high winds and sand the mission was scrubbed. I wanted to be a part of it and I wasn't. When I was told I was not going to be on the mission, I got a little frustrated because I felt like I should be a part of every mission that involved our team, particularly from a medic standpoint.

Going on those missions was a pride issue for me. I wanted to be there with my guys should anything go down. Sure, I had great faith in every one of the other medics who usually ran our patrols with us, but being there in the flesh was a different story. I would have eyes and hands on. Throughout the day and in most any situation I found myself in, I ran battle drills in my head. A battle drill is one's planned reaction to a given situation. I always ran scenarios through my head. The scenarios kept my mind fresh and alert to what I would do if something actually went down.

But today I just sat around all morning, checking my emails and watching *CSI: Miami*. After dinner I attended another medic recertification class at 2000. We learned a little bit about how to treat patients suffering from cardiac and respiratory trauma. It was a good class. The Colonel who taught it was a member of the Kansas National Guard. He was attached to us until May. He was very thorough in his presentation. The second class we had was how to treat dislocated joints and fractures of the bones. It was a good hands' on class. Though we did not have any real world injuries, we were able to work through simulations, applying bandages and splints to other medic's extremities.

This afternoon I did two tattoos on a soldier from the 386 Engineers. The 386th was attached to us from Texas. Winky got a touch up of an old bio-hazard tattoo he had gotten in Hattiesburg, MS just weeks before we left for Iraq. The actual bio-hazard sign was in black. The inside space had been filled with red. The red had not stayed well so he needed a touch up. He also got a new tattoo on his shoulder. Uncle Sam's Misguided Children was a tribute. He was never a Marine but he had a friend who was serving in the Corps in Iraq.

April 17

Yesterday our mission started at 0700. The XO, Apache Five Foot, briefed us, we loaded up, and then we headed to FOB Knot. Dark Wing, who usually rode as tactical commander of our vehicle, decided he wanted to drive so I took over as commander. It was not a big deal but it was a nice break from having to drive. In reality, our leaders should have made that happen more often. Too many times senior NCOs and officers ran everything and never allowed the junior NCOs to step up and gain any relevant experience. It happened all too often in our company.

We arrived at FOB Knot and were met by a platoon of IA from Charlie Company. Abas and LT Ali greeted us with warm smiles and hearty handshakes. The hand greeting in Iraq was a little different than it was in the U.S. If the person one was meeting wasn't someone already known, one didn't give him the ever loving embrace and cheek kisses that normally would be given in this country. The two people meeting, after shaking right hands, would then cover their hearts as if to say, "Peace be with you from my heart."

LT Ali's platoon was working with us today. Abas, along with our base translator Jack, were our interpreters. Jack was an interesting character. I say that because he was so animated. He was probably only five feet tall but he was as hairy as a bear and as nice as Mother Theresa. There was a lot of love stuffed into such a small package. He was a great guy, loved learning English, and went out of his way to assist our unit with anything he could do to make Iraq a better place for all of us. On this day we were going to oversee the shooting skills of the IA. It was an interesting morning.

For the amount of collective time that these guys had in combat they were not the best shots in the world. Out of over 50 soldiers, only five, ten at the most, could actually shoot well. All the others had bad technique. Technique could be trained out with time and with enough bullets. There were a few who were simply terrible at shooting and it was all of our hopes that should we ever have to come back to fight in Iraq, they would be the guys we fought. Easy win for the Gipper.

Some of the guys were not even remotely close to hitting the silhouettes that we had 25 meters down range. We spent a lot of time with them that day showing them how to breathe correctly when pulling the trigger and taking a single shot instead of relying on automatic bursts. They just needed more practice. When soldiers looked like they understood what we were showing them and they adjusted their fire and were consistent in their shooting, we would let them sit down and a new soldier would take their place and the process would start all over again.

The IA loved our equipment for posing purposes

I shot the world famous AK-47 for the first time. Because I had never shot one, it seemed to have a bit more of a kick to it than an M16 or M4. I still shot pretty accurately. I've heard that about the AK. If one takes his time with it, one can hit any target. Most of the people who fire the AK could be much more effective with it if they just took the time to line up their shots before they fired. It was safe to say that the majority of the time, when an AK was fired, it was fired on full auto so a "spray and pray" philosophy had been used throughout the long life of the weapon. It was still an excellent weapon despite the people who used it.

After shooting the AKs, we took some shots with our M4s and allowed some of the IA to shoot them as well. I liked working with the group of IA stationed in Balad Ruz. They were usually fun guys to be around. They laughed more than most of the people in Iraq that we saw on our patrols. Granted, we got to interact with them more often than with the general public, so they got to see us and we got to see them in their relaxed modes. I was thankful we got the opportunity to see that because Iraq was not a place where we saw a lot of hope.

April 20

I've gotten in the bad habit of waiting two or three days to write in my journal. Not good in the long run because I'm sure I'll forget some of the events and how they all happened. It was Wednesday night and the week had been moving at a breakneck pace, thank God. Being slammed with missions and work helped to make things here move more rapidly. My cousin had told me about that. He recommended that I get into a routine as soon as possible because it would make the time pass more quickly and I would not be looking for things to do to pass the time when I did have down time.

Sunday I went out with Lima Bean and his crew. I enjoyed going with them. They were a bit more military than what I'd been experiencing with the Deuce. I think most of that came from Lima Bean's stint in the Marine Corps. Our mission was slotted to go to White Castle, skirt the northern border of our AO, and then come south on Route C1 into Balad Ruz. As with most things in life, things changed often in the military as far as operational planning was concerned. Two or three of our MITT teams had been slotted to escort IA Recruits north to their basic training site. Lima Bean wanted to ensure that all the Hummers under his command were mechanically ready for the trip. He curtailed our patrol and we only went to White Castle before heading back home.

Our Iraqi advising team

On Monday morning I got up early and went to wake up Crowfoot so that we could go eat breakfast. After breakfast we rallied at Apache 33, our Hummer. Dark Wing and Pretty joined us for this trip. Nine Hummers would be making the trip along with five IA vehicles and ten buses with recruits. We got another brief and then headed to the staging area across the base where the Iraqis were waiting. Our convoy commander briefed them and we headed out a little before ten.

It was quite an amazing sight to behold. Those guys were the future of Iraq's national security! We were a part of their historic moment. We were given the opportunity to escort them to their training camp. One of the reasons it was incredible was because there had been reports of previous convoys such as ours which had gone from Point A to Point B and in the process had lost their full complement of recruits along the way because the guys were scared to death of doing their duty for their nation. The insurgency was that effective.

We had an open seat in our Hummer so we picked up a captain who was acting as an advisor with the Iraqi basic training. He was a nice guy and said he'd been here before in Operation Iraqi Freedom I. We asked him what the differences were now from then. He said that in the initial assault no one had to worry about IEDs. Now, they were so prevalent that IEDs were what one had to watch out for more often than any gunfight. Furthermore, he said that he'd come under attack, either by IED or small arms, five times already in his life. He chuckled a bit at his own words and knocked on wood (the Hummer door) as he did. We assured him that we would do our best to keep him from any unnecessary death or dismemberment.

Apache gun ships escorted us to Anaconda. Those were some awesome birds! We were not used to driving with them flying above us for security so I'm sure we became a little more complacent than we should have been on that particular part of the route. As soon as we got on Route T2 we ran across a cordoned IED being prepped to be blown. Because the perimeter had already been set, all we did was stop in route and pull security on the vehicles and IA Recruits we were with. We were stopped for about an hour before EOD finally blew the bomb so all the recruits were able to take breaks and get lunch.

The scenario was pretty ridiculous. An entire convoy was out in the open. If insurgents had wanted to light up some buses or military vehicles they could have done it relatively easily. Now, realize that as soon as they started shooting, they would have been lit up like the Fourth of July from the ground and air. The recruits kept getting off the buses and doing their business on the side of the road. But because of their Muslim beliefs, they could not piss standing up. They had to go far off the road and squat down to urinate. It was funny for us because it was so far out of our box. Finally the IED was blown. It was a small explosion but if a bus or Hummer had been hit by it, I'm sure some damage would have occurred.

The base we were headed to was called Speicher. It was the old Al Sahra Airstrip outside of Tikrit, Saddam's birthplace. On our way there we crossed the Tigris a few times. Tikrit was big. We were all chomping at the bit to check it out, but because of the hostile environment there and the precious cargo we were transporting, our leadership kept us away.

The base itself was pretty cool as far as bases in Iraq go. Since we had never been there before it was that much more interesting to us. There were not many soldiers on post but it seemed to be a big hub for convoy traffic. There were a lot of transients, both military and contractor. When we arrived, the whole post was going a little crazy because they had never hosted Iraqi Army soldiers. It was a norm in our part of the dirty little war. We worked with the IA every day. The Speicher CO almost did not let us stay, but after a careful conversation with C Major, their CO gave us the green light with our IA friends.

We were escorted everywhere we went: partly for security purposes and partly, we liked to think, because we were so important. The Base Reaction Force was a great group. They bent over backwards to ensure that our stay on Speicher was a good one. We were taken to dinner chow in one of the finest cafeterias I'd seen to date. We also got to utilize their PX, coffee shop, and souvenir shop. Our biggest concern was that the IA soldiers didn't have any money to buy things at the PX. Most of them sat outside the building waiting for their mates so we could all head to our sleeping quarters. I gave $10 to our interpreter, Jack, so that he could buy a new arm wallet.

After we shopped for a bit we were led to tent city where we stayed for the evening. We got settled in and of course all conversation went to women. One can guess who led it - that's right, Pretty. He boasted of his escapades as well as some of his big busts. I'd heard most of the stories a few times already but they were still pretty funny. Other Joes chimed in and soon most of the tent was rolling on the floor with laughter about past sexual relationships. There were some of us who tried to read but every once in a while a good story would slip in and we'd all chime in with laughter. The quote of the evening was from Shrek. He was trying to read and said, "Will ya'll shut the hell up? I'm trying to read my damn Bible."

The next morning we got up at 0700 and made our way to breakfast. It was a good experience in a great cafeteria. Not only was the food delicious but there were women to be seen all around us! I know what you're thinking, "These guys are animals." But one has to understand that we'd been in the desert for over five months at this point and away from our families for almost a year, so even the least beautiful women made the top ten lists. We had to have something to anticipate.

We left Speicher at 0900. I thanked the BRF for being so kind to us and they seemed more than happy that they were given the opportunity to help us out. It was good that appreciation went both ways. Sometimes competition compels two units not to treat each other with respect. In this case, we were trying to accomplish the same mission and with that group, it was no problem.

The drive back seemed to take forever and not much conversation went on. I popped a couple of Ripped Fuel capsules. I was more alert and more focused on what I was doing. We made it to Anaconda without incident. Rather than going on post, we decided to pull over outside the base for two minutes of roadside relief. As we drove past the Anaconda area of operation I thought about how good it would have been to have the Apaches flying with us again.

When we got near the Tigris River I asked Dark Wing to be ready to get an Iraqi knife for me. The last two times that I'd been past the river, kids had been out selling IA knives. They were not as good as American made blades but they made good gifts. A little bit of Iraq went a long way. The kid we did see was asking more than I was willing to pay so we drove past him.

From that point on we really had to start watching our gasoline. C Major's Hummer was running low so the word came over the radio that we were going to stop at FOB Warhorse for lunch and refueling. When we got there, the CO would not let us come on post without a big hassle because we had IA with us. We were already past the gate and at the clearing barrels, but C Major made the call and said, "Screw 'em." We headed further down the road to a point where it was safe enough for us to stop and refuel our vehicles with the two five-gallon jugs that we had each brought along. After refueling, we drove 30 more minutes to Caldwell.

The following day we did a normal mission schedule. We went to IA HQ and spent time with our Iraqi counterparts. Like I've said before, most of the time we were there we liked to interact and teach the IA, but as our time together went on we ended up hanging out because we only had one or two interpreters. At the end of our stay we were told that we would be going on a small raid because an informant had come in and ratted on his neighbors who had unauthorized weapons stored somewhere in their house.

Everyone was eager to roll. The Iraqis loaded up their trucks, slinging a ton of dust in the air as they hauled ass out of the HQ parking lot. We moved south to the location, stopping at two other houses first because the informant "wasn't sure" exactly which house it was. We moved to the last location and after tearing a hole in one wall of the house were able to find some AK-47s, a shotgun, a few other guns caked in rust preventative grease, and about 1000 thousand rounds of ammo. It was a good find and the IA did a good job of running the raid as we simply acted as advisors.

****

From: *<Malcolm>*
To: *<Becky>*
Sent: *Tuesday, April 19, 2005*

Hey,

Just got back from a two day trip up into the area of Saddam's hometown of Tikrit. Man, it was a long drive from where we were! We escorted a bunch of new Iraqi Army Recruits to the place where they would be doing Basic Training. The trip was cool in that we got to see a lot of stuff we hadn't seen yet, but it was just long and in military terms, it's always longer than it needs to be.

April 23

Today we were up at the crack of dawn for a 0530 rally at the Hummers. We were going to drop off Handy and a security element at FOB Knot. The rest of the team was headed into Balad Ruz so Apache 5 Foot could speak with the IA commander about training and upcoming exercises. Three gun trucks went to Balad Ruz. One stayed with Handy and his construction team.

Our time at the IA Headquarters seemed to drag. I did not want to get out of the Hummer because I knew that as soon as I did, the IA soldiers would start asking for stuff. Sure enough, as soon as they saw me they started asking for bug repellant, DVD players, and an assortment of other things. I liked trying to help them out, but at times it got overwhelming because even when I expected a little down time, there was always someone getting into my space wanting something.

Alone time in Iraq was next to nil. When on base, I was either in my room, in the computer room, at the gym, at chow, or buying a new movie. Everywhere I went, someone else was there. The only alone time I ever got was when I was in the shower or taking a dump and even then, guys were usually in the general vicinity. For some Joes that wasn't a big deal. They really enjoyed being with other people. I, on the other hand, missed being able to disappear for hours at a time.

After finishing at the IA HQ we escorted the Apache 5 Foot back to FOB Knot. Sixty miles per hour in a regular car was nothing in the States, but in a Hummer down Iraqi roads it was always an adventure. The potholes offered us opportunity to get airborne every 100 feet; and the brick trucks that ran the road, as well as the tractor drawn wagons, were good to weave in and out of. Every time we went outside the berm, we had as much to worry about with them as we did with bombs placed along the side of the road.

When we got to FOB Knot we quickly set a security perimeter. Everywhere we went and anytime we stopped, our first priority was to set a security perimeter. It was like clockwork. In reality, we tried to find a place where we were shadowed from the sun. In the Iraqi desert there was no place to hide from sunlight. The sun would hit anywhere we went, and if we did find shade, the hot air would remind us that we couldn't escape the heat. This week it had remained near 100 degrees and it wasn't even the hot spell from what we had been told.

The routine at Knot was usually the same every time we went. We would come in and pull security while Handy supervised the IA soldiers building a range. We always worked with IA while we were there so they could begin taking ownership. The FOB was going to be theirs just like the country. Better to let them start small was what we thought. It had really started taking shape. We tried to offer help but Handy liked to work by himself. However, he did utilize the IA and at times, even Cow Tits was allowed to help. The rest of us would just sit in the Hummer and read, talk, laugh, sleep, and sometimes actually pull security.

IA stacking during training

Two days earlier, on an IA led raid, we had confiscated some grenades in a cache search. Binky told us that they were smoke grenades. Big Red decided to allow one of our guys to throw one of them to see what it would do. Cow Tits, like any good baseball aficionado, reared back and threw the grenade over the berm. Having believed that it was a smoke grenade, I don't know why he threw it over the berm, but it was a good thing he did. The next thing we heard was an explosion. Turns out, it was not a smoke grenade at all. Fortunately, Cow Tits had thrown it far enough that no one was in danger of being hit by any debris or shrapnel. Joe with explosives, now that was a dangerous combination.

Dark Wing found some spray paint and tagged one of the Hesco Barriers. He renamed the FOB from Knot to Handy. When he was through, I took the paint and tagged a Combat Carpenter symbol beneath the words, "Welcome to FOB Handy." The Combat Carpenter symbol has a crossed hammer and screwdriver, surrounded by an olive wreath to recognize one's carpentry experience in combat. Although Handy had never come under fire while doing his construction and although this was a fictitious award, we felt that Handy deserved special recognition for all the work he'd been doing to assist our Iraqi brothers.

We finished our work at FOB Handy at 1230 and headed back to Caldwell. At lunch The Big Red One jokingly told me that my speedometer must be broken in my Hummer because he said we were doing 65 the whole way back. He usually liked going only about 45 to 50. I felt much safer going at higher speeds because of IEDs. One of the classes we were given while we were at Camp Shelby was on IEDs and was taught by two former Delta operators. Two of the main points I took from the class were drive the Hummer like I stole it and own the road when I'm on it. By that they meant, drive fast and drive in the middle, letting everyone who is not American know I mean business if they dare to come anywhere near my convoy. Don't let anyone break into the convoy and definitely don't be the guy who allows a vehicle-borne IED to detonate in the convoy because he wasn't aggressive or paying attention to his surroundings.

When I got back to the CHU I got ready to do some tattooing. The first guy to get a tattoo today was supposed to be Carlos, one of our interpreters. He was almost 17 and was in a line of work that most Iraqi men were scared to do because the reality of immanent death was in his face every day he chose to help coalition forces.

Carlos already had a girl's name tattooed on his arm but he wanted to cover it up with an American style tattoo. I was more than happy to do it. He was a good guy - very brave. He hadn't quite decided which tattoo he wanted to get so I did one on another guy, Smith, which only took about 30 minutes. Just as I finished Smith's ink the power went out for some scheduled cleaning of the generator. Carlos' attempt to get a tattoo was thwarted.

I suggested that Carlos come back after 2000. He did and got an eagle on his right bicep and some barbed wire around his left bicep. He did really well for a 16 year old. There was no crying or complaining. I liked that because it helped me get the tattoo done faster. I try to discourage young people Carlos' age from getting tattoos. Not only do I think somebody that young doesn't really know what he likes for a lifetime, but I think if a parent does allow his child to get one before he can make that decision on his own legally, that parent might just be a redneck.

****

From: <Malcolm>
To: <Mom>
Sent: Tuesday, April 26, 2005

My buds from the contractor company, KBR, came over yesterday and fixed my A/C in about 10 minutes. They said they saw it was mine and got there as quickly as possible since I am the tattoo man. Ha ha! I even started a new tat on one of the guys. He's real pleased with it so far. It's a Samoan style piece around his bicep. I can't wait until it's finished. It's going to look good.

So how are you guys doing with Loco gone? Becky said Dad was a little depressed. I hope he's all right. How's Black Dog handling it? Is he lonely? That's crazy. I thought Loco would outlive everyone.

Cobra Commander

May 8

I've been out of the loop for a few days but there has been a good reason. I've been on a mission to another base for the last ten days. As fate and good fortune would have it, I was "ordered" to go to FOB Cobra on a tattoo mission. Third Squadron of the 278th resides there and their commander requested that I be brought to base to do a few tattoos for some Iraqi comrades of his.

I left on Wednesday, April 27th, with every intention of coming back to Caldwell on the following Saturday. The demand for my work kept me there an additional week. The Colonel was really good about allowing me to stay for those extra days. He took the heat from 1st Squadron. It was great!

I arrived on Wednesday afternoon and was taken around the base by EM, whom I had known from my college days at Tennessee Tech. I hadn't seen her in ten years so it took me a while to recognize who she was. When I initially saw her I thought, "Hey. There's potential in this one. She's being pretty cool. Let's see what happens." She took me to the person I needed to see about checking in. I did and then caught dinner with Lug Nuts. First Sergeant Death Pants had quarters for me where I could also set up shop. I set up in the Iron Troop day room, a little room set aside for the soldiers in which to relax. It soon became overrun with Joes wanting tattoos and soldiers who were hanging out, waiting in line for their next piece of work.

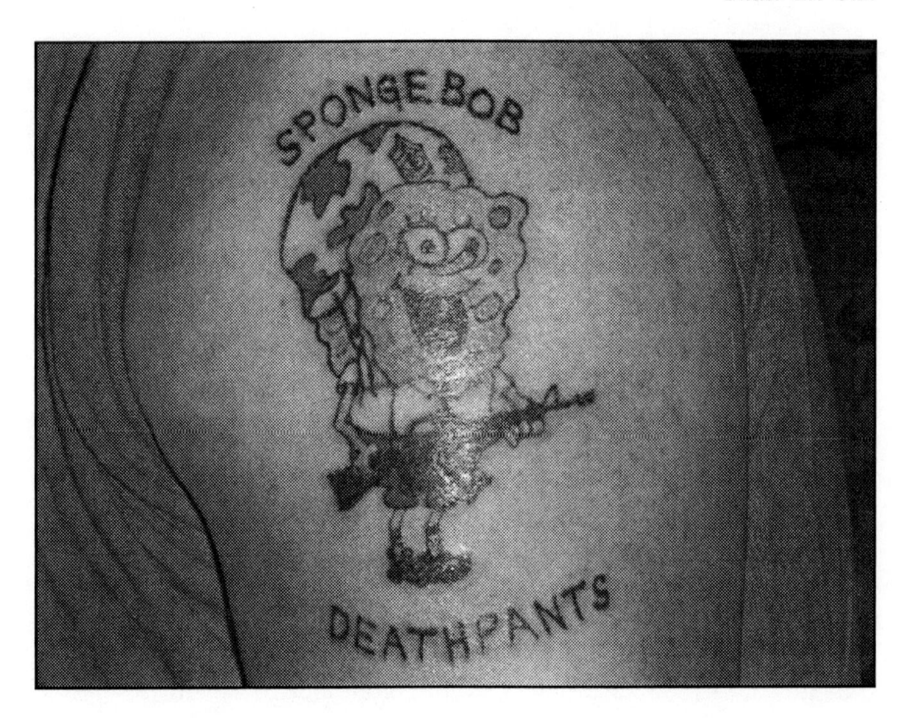

I started tattooing at 1800 that evening and didn't quit until 0200 the following morning. It was good times. Most of the guys who got tats that night had never had a tattoo and had decided that they wanted something as a reminder of their time in country. One of the big requests was the 278th banner. Its colors were red and white, cavalry colors, with gold crossed sabers and "Cavalry" written beneath. The Colonel had added the phrase "Freedom Isn't Free" above the flag. Most of the officers who got tattoos wanted that one.

Over the next few days the variety of tattoos was a relief for me. There were suns and moons, skulls, eagles, Pink Floyd album covers, crosses, and wives and children's names. Some were very low key but some were more memorable. One soldier's first tattoo was from the comic book "The Punisher." His platoon's nickname was the Punishers but he didn't want the Punisher skull that most of his platoon brothers already had. He wanted something a little more original. His friend had drawn a picture for him of a Punisher book cover. The face on the cover was half flesh and half skull with a ripped helmet. It was a good piece and a good challenge for me to do.

Another memorable first piece was on SSG Asbury Hawn. He originally wanted to get an eagle with Old Glory blended into his feathers but decided to get a Cavalry skull instead. The skull was all black work and wore the Cavalry Stetson. Beneath were crossed sabers and a banner that read "Cavalry." When I got through with it some of the guys said that it was the best piece I had done. I didn't know about that but I was pleased with it.

The tattoo I spent the most time on was of the POW / MIA flag centerpiece. I did that on Luke's leg. He endured four hours of tattooing to get the piece done. It was my favorite that week. Because a tattoo was like getting a major abrasion, he felt the pain for a few days. The night before I left Cobra he was hobbling around saying his leg still hurt. Ah, such was the life of the tattoo enthusiast! (I saw Luke a few weeks later, on his way home because of a hardship. His leg had gotten infected and he thought it stemmed from when a Hajji kid had kicked him in the tattooed area when we had gone out to eat during one of our breaks. He seemed to believe that the infection came from the dirt on the kid's feet. His was the only tattoo that year that I did that had any issues with infection.)

The highlight of my time there was the opportunity to tattoo one of the Colonel's guests, a General in the Iraqi Army. He wanted to show his appreciation and commitment to the Colonel by getting the 278th banner tattooed on his left shoulder. He had a good time with it but insisted that I not show any pictures of his experience to any other Iraqis for his own safety. I also put the flag on two of his interpreters, Al and Hang. They have been in country since OIF I serving the coalition forces. They were some of the hairiest guys I'd ever had the opportunity to tat. When I finished with the interpreters they invited me to their CHU for dinner. It was a good time hanging out with them and getting to know why they were doing what they were doing. They also had a good stash of whiskey. I didn't get hammered but on the way back to the day room I was feeling pretty good.

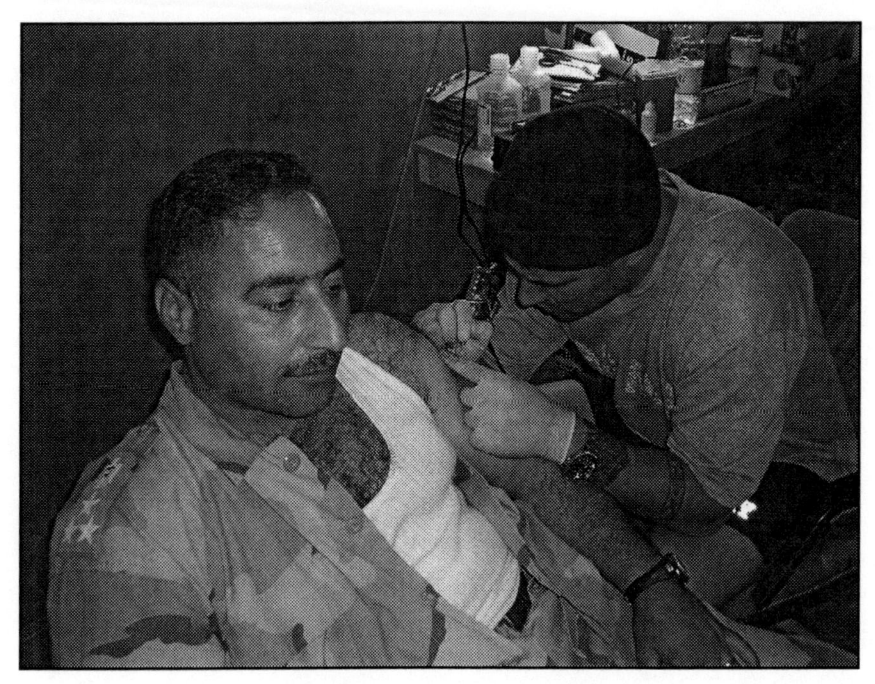

Tattooing the Iraqi General

Seven of the ten days I was there I was tattooing for at least 12 hours a day. I burned out two of my guns and ran out of black ink and needles. Fortunately for the guys and me another soldier on base was also doing tattoos. He was a beginner and was a bit slow from what I'd heard, but he was good. I bought some supplies from him and was able to continue my work. (The only other issue I dealt with at Cobra was that at some point the ink I was using went bad. Some of the tattoos that I had done faded after a few weeks. I had never had that happen to any of my work and didn't know if it was the ink or the environment. I suspected it was the environment because I'd been using the ink at Caldwell and it had done fine. Maybe it was the new ink.)

On Sunday, the guys brought in a tailor to have custom suits made for them at $80 a pop. A custom dress shirt was only $10 so I put in the order for a black suit and white shirt. I wanted to look like John Travolta and Samuel L. Jackson in *Pulp Fiction*. I couldn't wait to get it and see how it fit. I never did get that damn suit.

Two of the days I was there I was afforded the opportunity to go on patrol with Iron Troop's 3rd Platoon. Both days we ended up in the city of Khanaqin. The city was predominately Kurdish and was friendly to Americans. It was the best looking city that I'd been to in Iraq. There were actually a few attractive women there over the age of 15. That was a first. I was able to acquire some liquor there from a kid who apparently came around often when the guys patrolled the city. It wasn't bad liquor. It was out of Turkey.

Also while on patrol I got to see some oil fields. I was told that the oil comes out of the ground at the drilling site. I didn't think it would be a big deal but when I got there and saw just how much oil was coming out of the earth and lying around I was stunned. I thought, "Man. If only I could get rights to this land, I'd be a millionaire!" I suspected everyone thought like that the first time he saw that area. It also made me think about how high the gas prices were getting back home.

We also went to a remote city far into the bowels of the desert. While we were there our driver got our Hummer stuck in a river bed. We spent the better part of three hours trying to get it unstuck. Being there so long, a village elder even came out and helped us dig around the tires. It was a good experience working with the locals in that respect because they were laughing at our situation. None of us felt threatened while being there and they knew we were not there to mess with them. We were there to offer them assistance if they needed anything and we ended up being the ones who needed help.

Yes it's true, Hummers can get stuck!

On the second of my patrol days I got the opportunity to go to the Iranian boarder. It wasn't a big deal except that I was able to say that I had done it. Soldiers with automatic weapons were on both sides of the gate. The Iraqi soldiers looked much more relaxed. The Iranian soldier we encountered looked like he wanted to shoot us just so he could say he shot an American.

We also went to a few castles on the border that the Iraqi Army used as Border Patrol stations. The castles were not the type one would find in Northwestern Europe. These castles were small enough to be houses. Their exteriors were the only features that gave them the appearance of being castles. I have always wanted a log cabin to live in but after seeing those little castles I decided I would like to live in a castle. All the living quarters surrounded a large center room which had windows at the ceiling that allowed light to enter. The roof was accessible and would have been good for a deck or observation post.

<u>May 9</u>

This morning we headed out at 0630 for the IA HQ. We arrived to find that the guys we were going to train had been taken by Major Ali to Baquba. We didn't have anyone to train but Apache 5 Foot thought that we should still go to FOB Handy to do some live fire runs. A shooting range could be fun but this morning it just seemed to drag on until Dark Wing decided that he wanted to see what a fragmentation grenade would do to the interior of one of the shoot houses.

He was giggling like a little school boy who was about to pull the world's best prank on his teacher as we made sure he knew what he was doing. He stood about ten feet away from the target, pulled the pin, threw the grenade, and then began to run back to a safe area behind some barriers. The funny thing about it for us was that he wasn't even close to getting the grenade into the house. It landed on top of a set of Hesco barriers.

As he took off running we could hear him laughing hysterically. His run looked like Fred Flintstone trying to get his prehistoric car moving: the part where his feet move in place until he gets enough force to move the vehicle. That's what DW looked like. It was hilarious! We were rolling because of his poor throw, his cheesy running technique, and the thought that someone could really get hurt while we were out here doing something like this. As we waited for the grenade to go off we wondered if he had a clue as to how poorly he'd thrown the thing. It went off after several seconds and the blast was impressive for such a small explosion; but we hadn't found out what it would do in close quarters. Needless to say, we railed DW about his poor technique.

Pretty came up to where we were all still laughing at Dark Wing. DW handed Pretty a grenade and told him where he wanted him to throw it. His throw had not hit its target. Before we let Pretty throw his grenade we went to inspect where the first one had exploded. The crater wasn't too big compared to some of the IED craters we had seen thus far, but it did give us another good laugh.

Pretty got ready to do his business and we all moved away. He had a little trouble with the safety tape but was able to get it off eventually. He pulled the pin, launched the grenade into the house, and then turned to run. As he turned he almost lost his footing and did lose his helmet. We were all laughing at him. This explosion actually was impressive. The blast blew out two of the walls and sent rocks and shrapnel through some metal that was in the house. It was good. Pretty, for all the coolness he had, looked like he might have shit in his drawers when he lost his footing. Understandable.

Berry decided to brave the Iraqi bathroom on the FOB and stripped down to his boots and Under Armor shirt. While he was in there, some of the guys on his vehicle took his pants and underwear and hung them on the antennas of their Hummer. When he finished his business, he came out with a look of expectancy. He knew someone would mess with his stuff. Everyone knew it. That was the chance we took when we took a dump where other Joes could mess with us. He walked around half naked until he was able to get his pants off the vehicle.

May 12

The training of the Iraqi Army had begun. We were now the advisors to what would become either the backbone of a new Iraq or the laughing stock of the world. Initially, what we saw was a little disheartening. The structure of the units seemed very bad. They had very little support from higher up and their supply chains sucked at best. These guys were about to be a part of a machine that would probably have growing pains for years, and what was worse; their pay system was screwed up. Because of that, many guys would walk away from training.

Our advisor team had changed a little bit. We now had Shrek and Donkey working with us instead of Atlas. Donkey I didn't mind, but I could tell Shrek was going to be a trial by fire. Our routine with the Iraqis included PT, combat maneuvers, cordon and searches, and some medical training. Most of the medical training was done separately by the medics who had been slotted to do that teaching. For some reason, I wasn't pulled into that function even though they said they would utilize me in that role. I didn't get ugly about it, but I did wonder what the medics were up to and why they did not give me the opportunity in the beginning to be a part.

Donkey had been on emergency leave for a few weeks prior to this time. On his return trip to Caldwell, he ran into Little D at a base in Kuwait. Little D had left Caldwell almost 50 days prior saying that he'd hurt his back while on a patrol. We all knew it was BS. He was shamming in the back. I was surprised that Donkey was able to confront him about the situation. Donkey informed Dark Wing, via email, that he'd made contact. Dark Wing gave him the word to order Little D back to Caldwell on the next flight.

Knowing that his scam was up, Little D made his way back to Caldwell. He continued to try to milk the back issue as kidney stones but no one bought into his stories. He'd pissed everyone off. No one could count on him. Dark Wing was going to move him into my CHU since I was living alone again. Gunro had moved out to be with Deacon. I raised hell. I did not want another soldier who I had issues with moving into my CHU. I'd been putting up with crap like that since day one. I raised that point to Dark Wing a few times. He eventually conceded and tried to come up with another plan.

Little D did come to the CHU to drop off his stuff and I told him that he would not be smoking inside the CHU. He did not like that and said that I couldn't make that call. It became a pissing match and I told him I would win all the way around because at that point I didn't care about my rank anymore. If he wanted to keep playing stupid games with me I would pull rank; but to be honest, I just wanted to beat his ass. He ended up leaving that afternoon to sleep with some of the other guys.

Eventually we came up with an arrangement with which we could all agree. I would move out of the CHU and up to Dark Wing's room. He and the Big Boss Man shared a room. They had another roommate at one time but he had pulled the broken knee card so he could go home after he got a combat patch. Little D would move into my old CHU with a guy from Michigan who needed a bed. It all worked out for the best, at least in my case.

May 14

The day was spent on the firing range, training our IA Company. It went fairly well although our Iraqi brothers could not shoot worth crap. We were attempting to help them zero their sights. The shooter took three shots at a target. Where they hit would determine where his sights said he was shooting in relation to the center of the target. Once that group had been shot and recorded, another group of three rounds was shot. When that group had been recorded and compared to the first grouping, adjustments to the sights were made – either up or down and left or right. The next time the shooter shot, he would hit the target center mass.

That was an ideal situation. In the case of working with the Iraqi Army, things weren't ideal. We had a bunch of guys who had, more than likely, shot before but had not shot the way we were trying to teach them. They were used to a technique called "Spray and Pray." The Spray and Pray technique was one where the soldier put the weapon on full-auto and sprayed as many rounds as he could in the direction of the target. Some might hit but most would miss. We were trying to teach these guys to shoot one at a time and to make each shot count. It was a long, tedious process and one, when it was accomplished, would make them much better shots. Hopefully, it would help keep them alive a little longer in a firefight.

Our interpreter on this day was an Iraqi who had served in Saddam's Army but had most recently been a chemist at Baghdad University. Mohammed was an interesting man, who when asked, told us how to make IEDs. Talking to him made the time pass more quickly. I enjoyed picking his brain. He was very open about his past and why he was doing what he was doing. He also told us why Muslims hated

Americans. He said that since America always helped Israel no matter what, Muslims considered Americans Jewish. Muslims hated Jews. It was a cut and dry answer. I wanted to ask him why he worked for us now if he, being a Muslim, hated Jews.

Smoke doing an IV class

We stayed at the range for most of the day, taking breaks only to eat lunch and for me to go to medic recertification class. It was a dry class but one that was a backbone for medics, patient assessment, evaluation, and care. The main reason I kept attending the classes at the Troop Medical Clinic was for the opportunity to see the female medics. That was about the only time during the day that I encountered women on base. It was always nice to think I might see one of the better looking ones.

****

From: *<Malcolm>*
To: *<Mom>*
Sent: *Saturday, May 14, 2005*

The building I'm in now is the same one we were in originally when we got here; only [now] there are just three of us in the room which is awesome. It's got A/C and showers and toilets down the hall, which beats the heck out of walking across base to use them. I like it much better – better atmosphere.

May 17

Yesterday and today were spent at the range. It was all right because we finished early and did not have anything slotted for the afternoon. Yesterday was a little more active though. One guy in particular, Nazir, could not hit the broad side of a barn with any of his shots. His excuse was that he was shooting with a different rifle than the one with which he had zeroed. Joe was the same in every country. He made excuses and the NCOs were giving him a hard time because of his shooting and his attitude. It escalated to the point that the entire group that was shooting had to run to the fence line a few times because he couldn't hit his shots. Eventually, at the end of the session, he and another soldier we ordered to do "low crawl" drills along the shooting path, which was gravel. I had been taking pictures all day long and this training evolution was no exception. I took two photos of him and his buddy doing the low crawl. That's when he lost it.

Nazir began throwing rocks in my direction and saying something in Arabic that sounded as if he were complaining that an American was taking his picture and would end up showing it to the Iranians or something. Whatever he said, he was very upset and began to storm off the range. The Iraqi Sergeant Major tried to talk to him and get him to stay but he was hell bent on leaving the general vicinity. As he walked off, the NCOs all said he was a poor soldier and that they were going to recommend that he be fired from the Army.

So I caused an "international incident." Oh, well. I guess it was bad but I didn't lose any sleep over it. The only reason I included it here was because that was the most action we saw that day. After the range we headed back to our side of base and relaxed the rest of the day. I took a great two hour power nap, got some hot chow, and began doing a tattoo of lesbians doing a 69 on Super Dave's arm. It was a tribal piece around his bicep.

****

From: *<Malcolm>*
To: *<Bruce>*
Sent: *Thursday, May 19, 2005*

Hey Bruce,

Thanks for the encouragement. That "Dragon's Tail" Ride sounds like a good deal. I'm toying with the idea of getting a bike when I get home but also about buying some land. Maybe I can swing both.

Yeah, I'm ready to come home. This place is getting old fast. It's always over 100 now and because of it I can tell I'm not eating like I normally do. Not that that's a bad thing because I can still [stand to] shed a few pounds but it just sucks thinking that it's only May and we haven't even hit the pinnacle of heat. Oh well.

Hope you guys get to make it up to TN for the 4th. That'd be cool. I know Mom and Dad would enjoy having you guys back.

******

From: *<Edith>*
To: *<Malcolm>*
Sent: *Tuesday, May 24, 2005*

Hello Malcolm:

In honor of Memorial Day coming up, the Military Task Force here at CRMC gathered for lunch along with family members of service members. We had prayer and good fellowship. We are continuing to pray for your return home along with the others. We continue to remember the leaders of our nation as well. YES, we would love to hear from you. Add me to your [distribution] list. I don't know you but I look forward to meeting you one day. I work with your father – he is very nice. You take care and stay positive. Glad to hear you got our package! May God be with you.

****

From: <Malcolm>
To: <Becky>
Sent: Saturday, May 28, 2005

Nothing much new here. I just got done changing the shocks, springs, and wheels on a Hummer. Man, that was rough. Thankfully I had two of our company's mechanics working with me, showing me what to do. It went a lot faster than I thought it would.

I don't know what they'll do for Memorial Day here. Probably nothing but a better dinner. That's the way it goes. I wish I was at home celebrating it with friends. Oh, well. We've been here for six months. One day closer to getting home.

Under the Gun

One of the best things about being in Iraq was the opportunity I had to do a large number of tattoos. When I arrived at Camp Shelby in September of 2004, someone got the word out that I did tattoos. I'd been in business since 1992 but I hadn't brought any of my gear with me because I didn't know how the chain of command felt about that kind of thing going on. It wasn't until our flight out to the National Training Center in Fort Irwin, California that Terrible Teddy spread the word to our Squadron Commander and the Command Sergeant Major that I did tats. When they heard that I slung ink, was very professional, and was good at what I did, they called me to where they were sitting on the plane and asked me the usual questions. It was cool because when they finished grilling me they both told me to bring my gear back with me when we took our block leave after NTC.

Doing tattoos in the barracks wasn't the ideal setting to give a tattoo, but we made it as sanitary and comfortable as we possibly could. All my equipment was fresh: new needles, new tubes, new ink, you name it. The tattoo area was sanitized before and after each tattoo session. At that point though, men preparing to go to war could have cared less if the carcass of a dog was lying right beside them as they got their tattoo. The tattoo they were getting would define who they were for the rest of their lives, but more importantly, for the next year while they were in Iraq.

Normally we'd finish our day's work and then someone would head over to the PX to acquire some cold beer while I set up shop. The lighting wasn't the best in the barracks so we acquired an overhead light that could be clamped onto the bunk post nearby. It gave me sufficient light. I usually had a rolling chair to sit on and depending on the location of the tattoo I was giving, the client would either sit on another chair or lie down on a nearby bed.

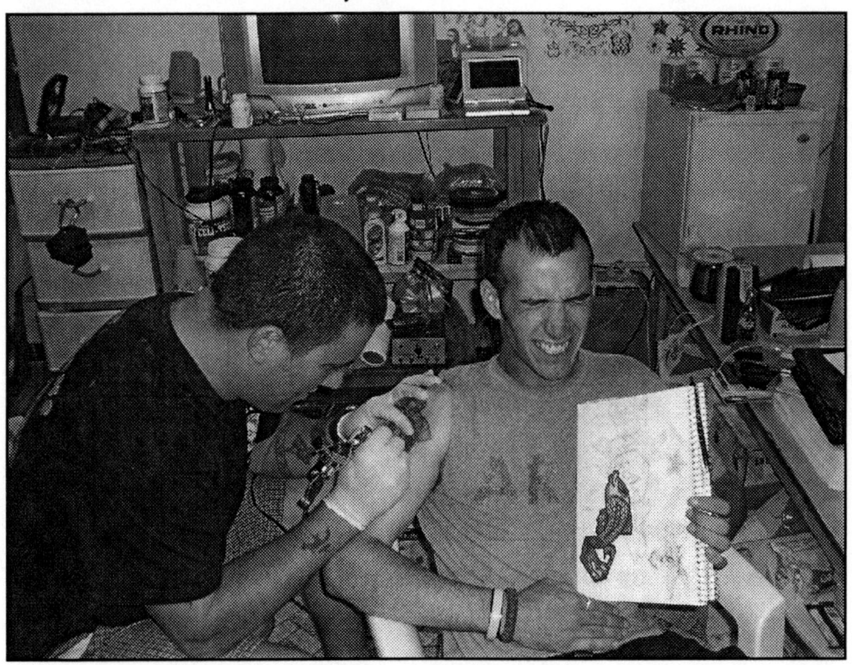

DT getting inked

The actual ink time became a sort of celebration for the guys who didn't have any means to get off post or who didn't want to spend the money to go off post. They would all, at one point in the evening or another, come through the barracks to check on the progress of the particular piece. Some would stay for only a few minutes and then come back throughout the night. Others would bring beer and hang out the entire evening. We had music playing, beer on ice, crazy stories, and enough folks around to harass even the toughest guy who was getting a tat. No man was spared. If you were in the chair you could rest assured that you were going to be the butt of all the jokes until the next guy got in the chair. That's just how it was. We had a good time.

A lot of guys got their first tattoo while we were at Camp Shelby. I enjoyed being the artist who got to hook them up, but I sometimes wondered when and if they came back from the tour, would they regret having gotten inked in the first place. Who was to know? It was a rite of passage. I wanted to get some ink done myself but I couldn't find anybody else on post who'd swap work with me. I sure as hell wasn't going to pay another artist $500 for work for which I could trade out.

On the capstone night of tattooing at Camp Shelby I had two girls scheduled to come in and get butterflies on their hips. One of girls had been coming by our platoon area and selling sex toys for the past few months. She had a really good sense of humor and wasn't scared to be around a bunch of horny guys, as cute as she was. She told me that her cousin and husband both wanted some work done that night as well. I told them to come over as early as possible because we had been ordered by Death Pants to have everyone not assigned to our barracks cleared out by 2300.

The afternoon started out uneventful enough. I began tattooing while the sun was still up. I'd sip the occasional beer along the way and folks would stop by as usual. As the afternoon turned to night though, the party cranked up quite a bit. Guys started coming back to the barracks from being out in town and were liquored up. More alcohol was brought into the barracks. Music was turned up louder. Movies were going in the more secluded parts of the barracks.

Then the girls showed up. As you could have imagined, everyone's hormones went through the roof as they walked through the door and down the hall. They both looked good. I felt sorry for the gal's husband because every Joe still around was undressing his wife. She was hot.

I started inking the girls and was able to finish them relatively quickly. Their butterflies were no bigger than the size of a dime. The husband was up next and we only had about an hour before every one had to be out. He wanted crossed Confederate flags on his left bicep. It wasn't a hard piece to do and he handled it well for his first tattoo.

While that was going on, Dark Wing and the LT had decided that they were going to box. Dark Wing had been in the barracks for some time getting liquored up on Jagermeister. The LT had been to an officer party hosted by the Colonel. He was hammered along with another LT who'd come to the barracks with him. Everybody was talking trash to everyone else, so of course, the gloves were brought out. Push came to shove and tempers flared. We tried to catch some of the fight but the only part I saw was when Dark Wing threw his last cheap shot after the LT had hit his head on some shelving. The LT was 50 pounds lighter than Dark Wing but you know what, they were both drunk as hell, so who really cared?

I finished the tattoo right as 2300 struck and the fight was ending. True to his word, Death Pants came through the barracks raising hell that all these extra people, especially girls, were still in the barracks. He ordered everyone out and all of us into our racks for lights out. He was ticked and we were all so pumped up from the night that we didn't care. We were all being smart asses to him as he turned out the lights. Crawfish sarcastically said, "Goodnight Top. We love you."

"Fuck you" was all he got out of Top. We laughed uncontrollably as he walked out the door. We laughed until we all passed out or fell asleep.

Tattooing in Kuwait was a bit of a challenge. We were crammed together in tents that held 100+ men each. I felt like I was a nomad and should have had some camels to ride. The tents had been doused in kerosene to keep the bugs away from the troops but in doing so the tents became highly flammable. Each tent had HVAC but none of them worked too well. One never knew if the air was going to be as cold as the air outside or even colder. It was miserable at times. The unit across from me exploded one night into flames. Fortunately, the guys who were sleeping in the vicinity acted quickly to smother the flame before the whole tent was ablaze.

As one might imagine, the worst parts of tattooing in a tent were the lack of light and the sanitation with all the foot traffic bringing in sand. Somehow I managed to keep my area relatively clean but the lighting was still an issue. After a few days into our stay, one of our

Sergeants broke out the overhead light that we'd used in Shelby. That did the trick until the day it fell and the bulb broke. We did not have access to anything over 40 watts.

I had a lot of business in Kuwait. It was still 3rd Squadron guys mainly because I had not yet been reassigned to 1st Squadron. Dark Wing, Death Pants, and the company brass were really cool about allowing me to do tats when I didn't have anything to do as far as work, live fire, MOUT training, or some other kind of training cycle. When Death Pants got his tattoo, Sponge Bob, five other First Sergeants came over to the tent to rag him about what he was getting put on his body. That was usually how it went.

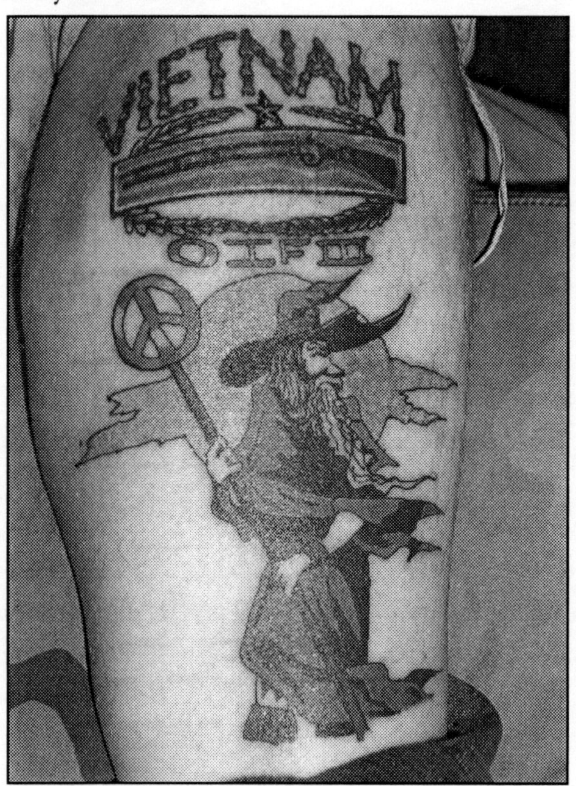

It was while we were in Kuwait that I realized how much potential there was for a good business while I was overseas. I'd done ink since 1992, the year I got off active duty from the Navy. I had a steady business over the next 12 years but nothing overwhelming. My work had been good enough though that people around my home town still called me up when they needed to have work done. But in Kuwait and ultimately Iraq, I would have a few thousand potential clients who were stuck in a central location and all they were doing was making money and counting down their time. If Joe wasn't on patrol or doing maintenance, he had down time. If Joe worked in an office on base, he or she would do his/her hours but would have down time.

It didn't take a rocket scientist to see that there could be some good money made as well as some really big pieces done. I'm glad to say that I did both. My only concern was the availability of supplies. I ran out of ink caps and some of my fine line needles while we were in Kuwait. Calling home to put in an order was inconvenient because we were nine hours ahead. I had to time the call just right to be able to reach Kaplan's in New York to place an order during their work hours. Since they didn't ship to APO addresses, I had to have the supplies shipped to my parent's house and they, in turn, shipped the supplies to me.

I only had to do that one other time while I was in Iraq. I made large orders so that I wouldn't have to try to get supplies often. My second order was from Superior out of Arizona. I had never used them before but wanted to try a new supplier. There was nothing wrong with Kaplan's services or products. As a matter of fact, they had been one of the best long time business partners I'd ever had the good fortune with which to work. I simply felt like I wanted to try something new.

It was also in Kuwait that my name started getting out to the other soldiers in our Regiment and to those who were attached to us from other units. The guys who were slotted to go to FOB Cobra were stoked because they thought I was going to be at their base the entire tour. Obviously, when I had to break the news to them that I wasn't going to be there at all, they were bummed out and kept trying to finagle ways to get me to stay with the unit.

It was amazing how Joes jockeyed for position in the tattoo line. Guys would ask me about getting a tattoo, but there were some who would always talk about getting inked and they'd never get in line or would come for a bit, leave, come back, and would expect the other Joes to have kept their place in line. Sorry, buddy. If Joe moved his meat, he definitely lost his seat. It was funny also to hear some of the conversations as guys tried to work their way up the ladder so that they could get inked sooner than their position in line afforded them. Rank was pulled. Buyouts were made. Friendships were tested. It was all in good, clean fun.

The tattoo that got the most attention while we were in Kuwait, other than Sponge Bob Death Pants, was the tattoo that Dark Wing got. Dark Wing was special. He always tried his damnedest to be the tough guy who outdid everybody else. This tattoo was no exception. All he talked about while we were in Shelby and Kuwait was getting a 9mm Beretta inked onto his chest with the words "Bow Down" beside it.

When he finally did sack up to get the tat, we had to shave him down with clippers because his chest hair was so full. It looked like he'd brought a rug along with him. After the shave, I laid him out on a black box that would serve as our work table. The tat began and within 10 minutes we all thought he was going to die with the way he was howling. Now granted, he finished it, but man, he was the biggest wimp I'd had under the gun to date. Every time I fired the needle up, he'd wail and cry about my going too deep. Getting inked on the chest was a tough spot especially since he didn't have enormous pecks that covered up his sternum. I'm sure the needle hit the bone more than a few times and more than a few times it was because I messed with him.

It was a funny experience for me because so many guys were watching him get it and having a good time. They were everywhere in the area that night. It was the biggest group of Joes we'd had to watch a tattoo since leaving Shelby. He was the Platoon daddy, so of course all his guys were going to watch and rag him throughout his experience.

When we did get finished and I'd cleaned him off and prepped him for healing, he swore that he'd never again get another tattoo. He looked like he'd been through the ringer and the ringer had won. He was in pain and did not appreciate the humor we were all enjoying at his expense. Oh well, it could have been worse. I could have made it last longer.

Initially when we arrived in Iraq I did not know how often I'd be able to do ink. We didn't have a set schedule for the first few weeks so I'd do a tat here and there. As the weeks gave way to months I was able to get into somewhat of a routine. My routine was this: anytime I wasn't on patrol, eating, sleeping, or working out, the shop was open. It took a bit of time for the word to spread around post but when it did, it was like wild fire.

I tattooed in my CHU for the first few months before I moved up to the Ivory Tower. It was a little cramped at times and when a client wanted something done under his arm, on her side, or in some other place where I would need that person to lie down, I'd have to rearrange a lot of things to make accommodations. But it worked well. It was a location that people had easy access to and when I wasn't around I could leave notes up to inform them when I'd be back.

The 386th Engineers were the clients who first came on strong. They were the guys from Texas. At least one of them would come by every other day looking to get some work done. That was one of the perks of doing the ink too. I was able to meet and become friends with a plethora of Joes whom I would probably have never taken the time to talk to had I not been doing the ink. Some really good combat friendships were struck up with X, Rios, Jr., Winky, and so many others just from the 386th. They were great guys who worked like dogs in Iraq. They were the guys who provided security for EOD on their missions and would blow caches that other units had found and confiscated.

One of the most intricate and largest tats I did in Iraq was on Vort. He came to me with an idea for a tribal piece with a dragon intertwined within the black work. I took his concept and drew it up on paper for him one day and one would have thought he was having a baby. He was really excited! We started the next day.

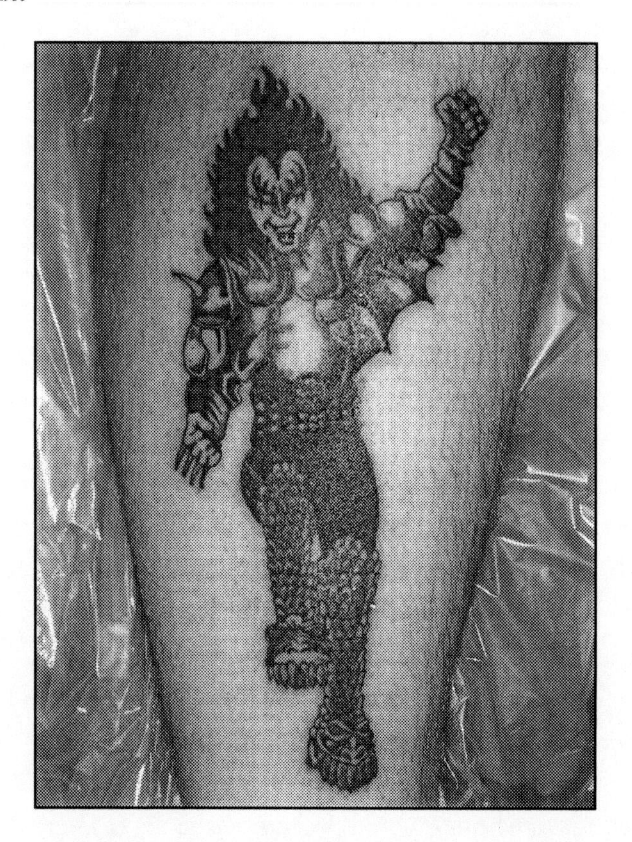

The upper arm was one of the best places to get inked because it was one of the least painful areas to get stuck. Vort's tat began an inch above his left elbow and wrapped around his shoulder onto the upper part of his chest. From the elbow to the shoulder he did all right. It was when I crested the shoulder and headed south that things got really ugly. He said that he'd never experienced pain quite like that. I could tell that he wasn't going to last past the outline that first day. He was thankful that we didn't go any further.

I usually had a really good time doing tattoos. Sometimes I thought the design choices were bad and I didn't enjoy doing them, but that didn't lessen my focus on them or the attention I gave the clients when they were under the gun. It just meant that I had to come up with ways to keep my mind sharp and the atmosphere light in the room. That's when I would start messing with the client. In Vort's case, I would go back to his upper chest area and stay there a few seconds longer than

I knew he wanted me to. I could see in his eyes and facial expressions that the pain was getting intense, so I'd just mess with him a little bit, staying in that sweet spot for a few more seconds. It got to be a running joke that he hated even though he'd bear with me when I did it.

One of the other things I started doing to make the time pass while a client was in the chair was play movies. Joe would come in to get ink and I'd find out which movie he or she hadn't seen yet and we'd throw it in and while I worked, everyone else in the room would get to watch a movie. Movies helped to keep the client's mind off the pain somewhat. We were able to buy movies dirt cheap in Iraq. Most were good quality but there were a few where we watched and listened to the people in the theater more than we caught the movie itself.

People seemed to really enjoy this aspect of the tattoo environment. When I moved into the big room, sometimes guys would stop by and hang out just to watch movies. They got tired of hanging out in their own CHUs and knew that they were always welcome to my room. Big Boss Man and Dark Wing didn't care unless they were taking naps and even then, it wasn't a big problem.

Tattooing in the Ivory Tower was great. It was a hell of a lot roomier than my old CHU and I was able to keep it much cleaner; that was a good thing when one was doing tattoos in a desert. It was by far the best setup I'd had to date since bringing my gear to Shelby. We had a few chairs, a table for lower back pieces and under biceps, a TV, and fridges to keep the needed Liger energy drinks cold so that I could keep my motivation up for the 13 hour days of tattooing that sometimes popped up.

I did tribal designs, cartoons, patriotic themes, and memorials to family, both past and present. There were as many different concepts as there were soldiers on post. The biggies were tribal though. I thought that fad would have burned out years ago but it seemed to be as strong as ever. An Air Force guy named Ollie, who became a good friend while he was with us, got two large tribal pieces on his biceps. They were the first tats he'd ever gotten. He went all out.

I was glad I had the opportunity to meet Ollie. His attitude and perspective on things in Iraq was so different from the soldiers who were on post. He and one other fly boy were attached to our unit for four months as forward observers. Their job was to call in air strikes should any of our units ever come under fire. From what he told me, he was bored to death on Caldwell because he and his mate weren't being utilized for anything except to stand one watch a day in the Regiment TOC. Not what he'd expected but he maintained one of the best attitudes I saw the entire time I was in country.

Some of the other guys who made things fun were the contractors who initially came by for tattoos, but after we became friends, they would just stop by to hang out, play guitar, grab a drink or snack, and then head back out into the blistering sun. Most of the contractors had been prior service. They said that's why they were able to stay longer than most people who signed up to work in Iraq and Afghanistan. They'd been through deployments before and they understood the needs of the military. They'd been there and done that.

Most of the guys got paid large sums of money for being overseas with us but they were very humble and would tip me very generously. They'd also put me on top of any work order lists I might have submitted work requests for because they knew if they hooked me up, I'd return the favor on their tattoos. It was a great relationship and would have been anyway had we not offered anything to anyone. They were just great guys: Mike, Dave, Robert, Jason, and all the others who came through. They, too, were a welcome break from the guys in uniform. When they came in, I got to hear what was going on with them, not about what our last patrol had been like.

Some of the guys who came to me for ink worked for security companies and had been contracted to train the Iraqi Police. They were pretty cool and I could always count on them to hook me up with beer or liquor for their tattoos. They had access to it and I did not, so it was a good trade. We were very discreet in our consumption, and as for the guys who were able to reap the rewards of those generous deposits, we thanked the donors for being willing to give us a hook up. Sometimes we just needed a break from reality, not to be stupid but simply to let our hair down and relax a bit, especially in that environment.

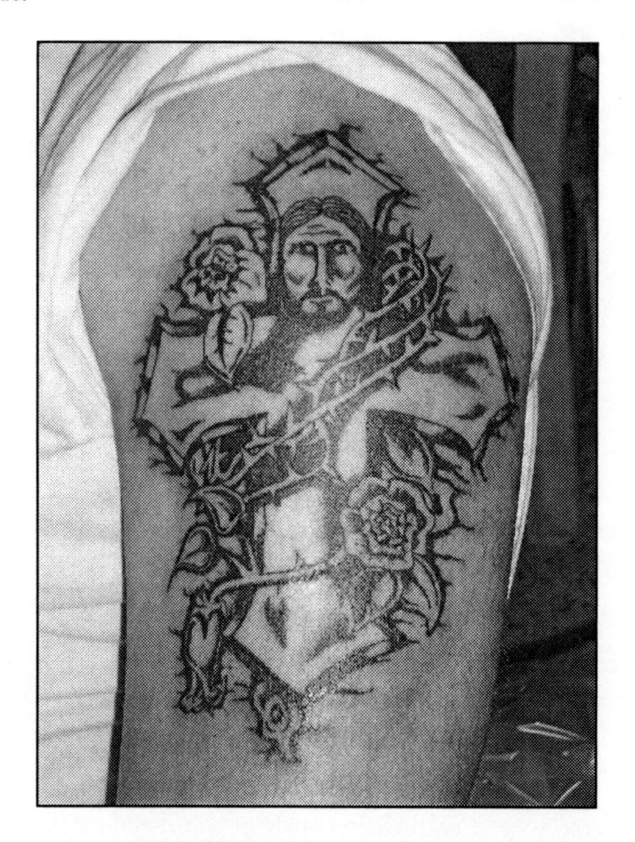

Some of the highlights of the tattooing included inking a General in the Iraqi Army, the COs of our Regiment and 1st Squadron, four Iraqi interpreters, and a plethora of officers and enlisted men and women serving with the 278th. The trip I made to FOB Cobra for 10 days of tattooing in May was incredible. There were so many great times hanging out and getting to know guys better. I would work for hours and Joes would bring me grub from the chow hall so that I wouldn't have to stop what I was doing. It was by far the best thing I could have done with my free time while I was in country.

Not only did tattooing give me something productive to do with my time but it lifted the morale of a lot of soldiers, mainly the guys who got tats but also quite a few more who simply hung out with us. There were two or three other guys on our post who were slinging ink as well, but from everything I was told about their work, no one got the full tattoo experience that I was able to give. I believed I was good, relatively fast, and was by far the most reasonable in pricing for the work that was being done. The way I saw it, I didn't have any real overhead. Sure, my supplies cost a bit but I could recoup those costs pretty easily so I didn't rape anybody on price. I think the guys really appreciated that too because they would come by time and time again. Guys usually didn't stop with one tattoo. One tat just wasn't enough.

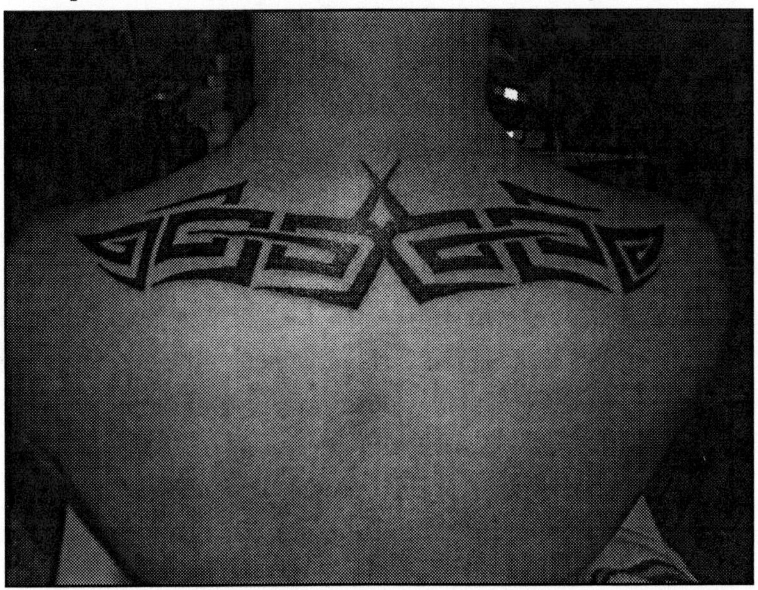

One of the weirdest locations for a tat was inking "Crazy" on the back of Crazy's head. We did it a few times to make it stand out more and more each time. He eventually came away with the letters done in red and shadowed in black. It wasn't anything spectacular other than it was on the back of his head and made people stop in their tracks when they noticed it.

Percolator got some of his ankle tat work given to him for free. He had been demoted and used that to work in a deal to get some money off the work I had done for him that night. I thought about it for a few minutes and then told him that if he licked up the blood that was on the plastic where his leg had been while getting the tat, I'd give him the work for free. He thought about it for a second and then bent over and began lapping his blood and ink up. It was hilarious! I didn't think he'd actually do it but sure enough, there in front of God and everybody (actually just Big Boss Man and me) he licked that mess up. Ah, the joy of dope deals!

Percolator was also the brave soul who allowed Crowfoot to do some fill-in work on his ankle piece. Old Crowfoot had been hanging around that night and my back was getting tired from having been bent over for so long. I made mention of it and he asked if he could do some of the fill-in ink. I looked at Percolator and he said, "Sure, as long as he doesn't fuck it up." I told him that I'd stand over Crowfoot to make sure he was doing what he needed to be doing. We switched off for the next hour. Crowfoot did a great job and really seemed to enjoy doing the work. He had some great tats on his ribs so I'm sure he felt a kinship with the work even more so now that he'd taken the opportunity to put ink in someone else.

One soldier inspired me to draw quite a bit after spending some time with him. DT had been drawing the entire time we were in Iraq, keeping a scrapbook for his wife and child. Some of the stuff was really good. A few of the pieces were intriguing enough for me to begin some works of my own in the style that he had used. He saw one that I'd drawn of a fish jumping in front of the sun. He liked it so much that he chose it to have inked on his right shoulder. That tat was one of my favorites. His friendship sparked a friendship with an entire platoon of guys - the COLT team. They were a lot of fun. I inked most of them.

Tattooing our commanding officers was interesting. The SCO came up to my room and got the 278th banner on his left shoulder. He did it very discreetly but everyone in the Squadron knew he was getting it. For some reason, no one else came in the room while he was getting inked. As for the RCO, he decided that he did not want anyone to be in the same area. I loaded up my equipment and went to his office. It wasn't a bad deal. It was the first time I had actually spent time with him. He had a flat screen TV on his wall so he threw in *Days of Thunder* and we hammered out his tat. Part of my payment from the RCO was a Regiment Coin. I'd gotten one from him already at NTC for handling a mass casualty situation during a training evolution, but I didn't think he remembered.

Finally, I need to tell a little bit about the girls who got tattoos. The best one was Rita. She was an attractive lady whom I'd met at NTC. A few of us thought that she was the hottest female her age on the base. She was only in her early 40's but she looked better than most girls half her age on post. I walked her up to my room one afternoon and gave her a rose vine on her lower back. She couldn't handle the pain very well so I broke out some whiskey for her. By the end of her tat, she'd drunk the pint. She was tipsy too, so I got a little make out session when we were through - very nice.

Outside the Wire

I sold my computer to one of our interpreters, Abas. I was not able to keep a good journal for the next two months. During the summer months we worked with the Iraqi Army, advising them in military matters and going on patrol with them. The guys we worked with were really good men. We enjoyed working with them and felt like we all accomplished something while we were doing that.

****

From: *<Malcolm>*
To: *<Mom>*
Sent: *Friday, June 10, 2005*

Hey! Guess what? The other two boxes arrived yesterday and I got them today. No problems. I'm glad they finally got here because the people were starting to hassle me about getting work done and I was running out of things to tell them about the mail system. Thanks for checking in to the USPS for me.

And, I've looked into doing something through the PX for a computer but they are always out on the base and [the inventory that they] usually restock from is out as well so it would be just as simple or more so to do it the way I've asked you guys to help me with it. Dad has the specs for the computer I want.

I haven't sent out the June update because I don't have a computer to keep my journal on right now. That's a big factor but I'm going to try to borrow someone's so I can type it out, save it on a jump drive, and then send it out on the emails.

****

Most of our advising with the Iraqis consisted of observing their training evolutions, firing range instruction, tower watch oversight, and quick response force for the patrols that we didn't actually go with outside the wire. We worked with a company. Some of the men had fought in various battles around Iraq. Some were very well seasoned soldiers. Others were fresh boots. Most were there to provide for their families' survival.

For their training evolutions, we would sit within hearing distance and ensure that a class was being taught. Now, unless we had an interpreter with us, we had no clue as to what the instructor was saying to his men. Most of the time we asked that they provide us with a schedule for upcoming classes and a sign-in roster for all men present. On the firing range we attempted to show the men how to zero and shoot effectively. That was the hardest task of all. Most couldn't shoot. Tower watch was a joke for us. We did rounds of the tower over-watch positions each night our company had the duty. Basically all we did was make sure that the IA stayed awake during their watch. They could hear us coming and all the towers had hand held radios, so I was certain that they woke each other up right before we made it to their positions. As for patrols, patrols were patrols. They were hunting trips.

Big Red and Jet Ski on patrol with IA

****

From: *<Malcolm>*
To: *<Mom>*
Sent: *Thursday, June 23, 2005*

No sickness as of yet except for being sick of being here at times. You know how that goes. I had the squirts for about two days when I was up at Cobra but that's been about it. I drink a lot of water and stay hydrated because it gets so dang hot out here. When I don't have to be outside I'm not.

Got a mission tonight. Should go all right.

****

From: *<Malcolm>*
To: *<Distribution List>*
Sent: *Saturday, June 25, 2005*

Hey folks. So I'm three weeks late in writing this. Oh well. Guess that's what happens when you have to upgrade your laptop and it takes forever to get a new one in. At least that's the excuse I'm going to use.

I'm going to make this short because I haven't been able to sit down and do my normal Journal Entry type update. For the past 25 days we've been training Iraqis. It's been interesting (sometimes) and quite boring at others. We actually act as advisors for the platoons. The sergeants in the platoons, who all have combat experience (particularly from Fallujah), train their own guys and we just observe and add any corrections we feel that are needed. I keep telling myself that this is a very crucial role in our getting home but sometimes the boredom in it weighs us all down.

The times we actually do get to interact with the Iraqi Army are usually when we go out on patrol with them. That's always good. Fortunately, the area we're in is not a Baghdad or Fallujah. We don't have the activity they do but there is still always the threat of ambushes and IED's. We never let our guard down outside the wire. I've come to realize that I really do enjoy going out on patrol. It gets me off the base and it affords me the opportunity to see some of the countryside, even if I've seen it a hundred times, as well as keeps me focused on why we're here in the first place.

The biggest thing going on right now is the "Mid-Tour Blues." Everyone's ready to go home. Some people handle it better than others. We've had quite a few guys make it home on leave and a few who are trying to play the crazy card to go home for good. Every one of us has issues back home but it frustrates me when I see guys manipulate the system so that they can renege on an obligation that they made to themselves, their families, and their country. Granted, some situations are legit. For instance, one of our guys, Scooter, went home on emergency leave because his wife was having a baby. The baby was delivered but died a few days later due to complications during the pregnancy. Fortunately, Scooter was home for that but he had to come back. To make a long story short, he tried to get a hardship discharge due to his family situation. The powers that be kept him here for a couple of months. He battled depression and anger during those two months, but was finally sent home a few days ago. Some of you had been praying for him. I appreciate that. He's only 20 and still has a world of growing up to do. That situation just made it harder for him.

On the lighter side of things, one of the soldiers, Goose, who was injured in an IED explosion a few months earlier and lost his foot, will be running a marathon with his wife within the week. How awesome is that! I can't wait to hear how he did. Apparently, he's a great guy with a great attitude and all his guys miss him back here and are rooting for him like you'd expect.

I've also got a thriving tattoo business on the side. It's been fun. I am usually doing at least one tattoo a day now. They come out of the wood work. Most of the work requested is patriotic. I've been able to talk some of the guys into letting me have freedom in the work once they tell me the idea of what they want. I'm looking forward to sharing some of the finished pieces with you all.

Okay, this is the last bit. I want to say a big CONGRATS to my cousin Stephanie and her husband, Al, on the new addition to their family - Jason Scott (I hope I got his name right - ha ha).

Thanks again for all your support and I truly am sorry for keeping you all waiting so long for an update. Thanks for all the packages you've sent - the cards, the emails, the prayers, and the thoughts.

****

From: <*Malcolm*>
To: <*Becky*><*Mom*><*Dad*>
Sent: *Saturday, June 25, 2005*

Anywho, yeah, I'm drawing up some new stuff for tats. It's been pretty cool. I've met a guy here who likes to draw but doesn't think he's very good. He is and he's given me some ideas from the things he's drawn. It's good. Maybe I'll take some pix and send them later on when I get a computer and am able to download and stuff like that.

Hope you guys all have a good time hanging out over the 4th. Wish I could be there. Won't be the same here even though they've got some event planned. But maybe it will mean a little bit more this year since we're over here fighting for freedom for the Iraqi people. Who knows?

****

From: *<Malcolm>*
To: *<Mom>*
Sent: *Saturday, July 02, 2005*

Sounds like you all are having a good time. Really wish I could be there. It was over 120 today. I'm sure it's hot [back home] but it would have been nicer [than being here]. I don't even feel like leaving the building on days like [today]. Fortunately I didn't have to [except for chow]. I did some tattoos already and should be doing some more tonight. I just got done with dinner and am headed back to the room. They're having boxing matches here tonight. Don't know if I'll make it with the tats going on.

****

From: *<Malcolm>*
To: *<Mom>*
Sent: *Wednesday, July 13, 2005*

I miss home. There's too much negativity here.

Could you send me a black berry cobbler??? Just kidding.

****

From: *<Malcolm>*
To: *<Becky><Dad><Mom>*
Sent: *Sunday, July 17, 2005*

Well Beck, where did you ladies end up going? Must be nice to have a selection. Ha ha. I chose not to go to lunch today because it's so hot outside and I hate walking down to the mess hall and not really being enthused about the food anyway. It's gotten monotonous (however you spell that one). We're thankful for what we have but sometimes it just gets old. You know what I mean.

Other than that, things are good. I might do some tattooing today but you never know with our schedule. It's crazy.

****

From: <Malcolm>
To: <Mom><Becky>
Sent: Saturday, July 23, 2005

Hey ya'll.

Good hearing from both of you today. Well, Cousin Mike suggested that I rethink my trip to England so I guess it's in the cards for me to take the trip home. I'm slotted for August 30. It may be one or two days here or there. Do me a favor and please don't make a lot of plans as I just want to relax when I get home and hang out with you all and hit a few rappelling sites. I'm sure I'll see folks but most of my time for doing that will be when I get home for good.

So that's about it for now. Everything's good today. Did a few tattoos. Getting ready to go back up to the room and take care of a few things before I get to bed.

****

From: <Malcolm>
To: <Dad><Becky><Mom>
Sent: Sunday, July 31, 2005

Hey Señor and the rest of the family,

I just got back in from a patrol with the Iraqis. It wasn't too bad. It was a little longer than expected but still not bad. It's almost 2200 here and it's still in the 100's. Crazy! Makes me have a new appreciation for the heat.

Things are fine. Sorry I haven't written much lately. I've been swamped between patrols, tattooing and some good workouts. Things are hectic but I was able to catch a nap today which was real nice. I about pooped out on our morning patrol. Oh, well, what can you do?

Can't wait to get home! I had to go see the Chaplain as part of the process for getting out of here. He said to be aware that things have changed back home just as much as I've changed by being over here for a year. I have to watch how I handle things when I get home because he said the stress levels in returning soldiers varies but there definitely is one just because of the change of environment. So I just thought I'd let you know up front if anything seems weird or I get short – please, just bear with me. I hope I don't [wig out] but you never can tell. I just want to relax. Slow down. Cool.

Well, I love you all and can't wait until the end of the month.

Go Rest High on that Mountain

From: *<Malcolm>*
To: *<Mom>*
Sent: *Tuesday, August 02, 2005*

Here's the deal. I get leave on the 30th. Now, from what I've heard, I'll head out a couple of days before and depending on the travel and slots on the planes, I'll either make it there on the 30th or a day or two before or after. So it's still kind of iffy. Like I said before, I'll let you know as soon as I start to head out.

As for Becky being there, that's a double edged sword. Sure I'd like her to be there but I also just want to relax. I don't want to be under anybody's schedules or high expectations of anything. Do you know what I mean? Personally, I'd rather you save the ticket for her for when I get home for good. That way I won't be rushed in seeing her. I think she'll understand. Besides, it's just a two week stay. In another month and a half from that we should be home for good. That'd just be better if you ask me … which you are.

****

From: <Malcolm>
To: <Dad><Mom><Becky>
Sent: Thursday, August 11, 2005

Hey Señor,

How are things going with you? Not too bad here. Kind of slow actually. I've been nursing an aching back for the last 36 hours. Too much time tattooing I guess plus wearing body armor and riding around in Hummers. Oh well. Could be much worse.

Keeping busy? Hope so. Look forward to seeing you in a few weeks.

****

From: <Malcolm>
To: <Mom><Becky>
Sent: Monday, August 15, 2005

Yo yo yo females!

Sorry I haven't written in a few days. We had four guys hit by an IED two days ago. Three of them died and the other, Downs, is in Germany, critical care. They were all in our platoon. Sad. Please keep our guys, their families, and Downs in particular, in prayer. Thanks.

Other than that, things are going as normal. We've got a mission tonight and tomorrow we'll head to the memorial service.

August 17 Update

Well folks, sorry again for taking so long to give you a journal update. No excuse except that a lot of what we've been doing over the past two months has been routine and, quite honestly, boring as hell. We've been working with the Iraqi Army, training them to be a better, self-sufficient team. How do they look? Well, let's just say that I would be very confident that we could still come back and take over Iraq if and when needed. We're without a doubt a much superior Army. Not that their training was sub-par but most of the IA were in it solely out of a need to make more money for their families. They were not so patriotic that they stayed to make Iraq a better place for all their people. They simply wanted to survive.

As for what's been going on lately, I'll tell you what's gone on since Saturday night. I had just gotten in my new computer so I stayed up late to mess around on it and play on the Internet. I stayed up until about 0100. About an hour later Pretty and Tater came into the room and woke me up with the news that three, possibly four, of our platoon brothers had been killed by an IED. Besides the initial wave of shock, all I could feel was sadness. Obviously, when you hear about a friend dying you wonder exactly how it happened and what the circumstances surrounding it were. But in a combat zone you have a tendency to go through those emotions rather quickly to a state of anger. Anger surfaced because some son of a bitch premeditated our friends' deaths by putting four artillery rounds underneath a culvert on a road our guys would use. The sons of bitches knew that our troops would go down that particular road in response to mortar attacks on their base. They knew that the detonation would not only destroy a vehicle but kill all the occupants in that vehicle.

From what we'd been able to gather, our guys responded to mortar fire on their base, FOB Bernstein. They were part of what was called a base reaction force or quick reaction force. When any issues came up involving security on the base, their platoon responded. The biggest problem we had when leaving either of our bases was that there was only one way in and out, through a traffic control point. Anyone could stand by, watch our routines, and decipher how we responded to certain situations. That's what happened with our guys on this night.

The team responded to the attack. The lead vehicle turned down a gravel road about 400 meters from the front gate of the base in the direction from which they believed the mortars to be fired. The insurgents had planned for this to happen. Somewhere along the road was a culvert and that's where they had placed the IED. The detonation device that was used was hidden inside what appeared to be a discarded garden hose. When the Hummer ran over it, the IED detonated underneath the vehicle.

Hummers with up-armor were designed to protect soldiers from blasts and shrapnel angled at the front and the side, but from underneath it was a different ballgame. The vehicle was tossed forty feet in the air and immediately became a fireball. Fortunately for the gunner, the blast threw him out of the vehicle, though no one knew that at the time. Everyone else on the patrol assumed he was still in the vehicle. Both of his legs were crushed as was his left arm. The fireball burned off his uniform and singed his hair and skin but he survived. His teammates found him a short time after the blast lying in a ditch some yards away from the kill zone.

The other soldiers who were behind the lead vehicle on the patrol were all members of our original Iron Troop platoon as well. They had the surreal experience of watching this violence happen and were helpless to render any aid as ammunition and ordinance inside the vehicle was cooking off. For a few hours they stood by as the fire burned itself out. There was nothing that they could do. Hopefully, our buddies died instantly as a result of the initial explosion. God forbid they were burned alive inside the Hummer. We all prayed that their transition was as painless as possible.

Yesterday we were given the gift of being able to travel to our brothers' memorial service at Bernstein. When we were in Camp Shelby we were told that one of the platoons from our company was going to be split up to help the mission once we got to Iraq. It was a sad parting but we knew it would only be for a short time and then we'd be back together again when we got home.

Memorial for Hawn, Taylor, & Reece

I'd often heard it said that death was a part of life. I agreed with that but it was the way a person died that made me reflect more about my own mortality and the way I lived life or my lack of living. Every time a soldier left "the wire" he or she put his or her life at risk. We all knew and understood this. All the military knew this and we did it anyway. It was part of the job. It was part of a life that we chose to lead. Soldiers, Marines, Airmen, and Sailors were inherent risk takers by the very nature of our chosen professions. Granted, some of us were more than others, but in the grand scheme, we all were. That said, our brothers knew the risks that were outside the wire every time they left Bernstein. They did their job. They looked straight into the face of that threat and said, "My teammates will be safe tonight because I'm going to do my job and I'm going to do it well."

SSG Asbury Hawn, SGT Shannon Taylor, and SPC Gary Lee Reece died that night doing exactly what they were charged to do. PFC Downs, the vehicles gunner, suffered catastrophic physical and mental injury. They were defending a nation. More importantly, they were defending their base and their brothers who were on it. SSG Hawn left behind a wife and four kids. SGT Taylor left behind his family as did SPC Reece. Even more than that, they left behind a plethora of men, women, and kids that their lives had touched and influenced, not only Americans who were at home and serving with them in Iraq, but scores of Iraqis whom they'd gotten to know over the past few months.

It was said at the memorial service that SPC Reece was a friend to hundreds of Iraqis in the neighboring city of Tuz. After the service, we went to our platoon's living quarters and watched videos of some of their crew's patrols. We saw many images of Gary Lee, one where he was presented with gifts from the locals simply because he took the time to get interested in their lives. He had a big heart and would always light up the room when he entered it. You always felt better after an encounter with him.

SGT Taylor was said to have been an outstanding member of the team and would always volunteer to assist with any job, big or small. His teammates said that Shannon could always be found during his downtime in the middle of the bunker room just watching TV and waiting with a smile on his face for someone to come in so he could give it to them.

SSG Hawn had recently returned from leave and told the guys that he planned to adopt his wife's two little boys. Asbury's heart was always going out to the guys, assisting them with routine, mundane tasks that most people would pass on. He was their go-to guy on computers and technical assistance. He was always pulling practical jokes on people.

Memorial for Hawn, Taylor, & Reece

We spent most of the afternoon with our guys from Bernstein. It was a good reunion under bad circumstances. I appreciated our Chain of Command for allowing us to go and be with them. I continued praying for them, particularly the guys who witnessed the ordeal. I couldn't imagine having to watch friends die under those circumstances. I didn't want to imagine it and I prayed that they would be able to overcome the nightmares as quickly and as wholly as possible.

Our convoy left Bernstein at dusk. It was an eerie ride back to Caldwell, the most tense I had been since getting in country. We all had bombs and death on our minds. We just didn't want another situation like that to happen. I was driving the tail vehicle and prayed the whole way back that nothing would happen. As I drove I asked God for our protection and that I wouldn't see a flash of light ahead of us from one of the other vehicles hitting an IED. We drove with purpose that night: bring everyone home alive.

The only break in the mood we had that day was that after we ate dinner at Bernstein we acquired a few cases of Beck non-alcohol beer. We loaded one case down on ice. On the return trip Top, Crowfoot, Tron, and I drank heavily. We tried our best to celebrate the lives of our guys. It was all we could do to lighten the mood. It was almost like we were in Tennessee on some back road, only we weren't.

Five days have gone by and I just finished this writing. It still seemed surreal. Those guys couldn't be gone - not this close to going home. But they were. And we must "endeavor to persevere." We would.

Soldier grieving for KIA

AUGUST 20

I got up at 0700 and walked across base to the chow hall for breakfast. I had the usual: three fried eggs, two boiled eggs, bacon, some French toast, and cereal. It wasn't too shabby. In fact, breakfast was about the only meal I looked forward to eating. I didn't lose much weight in Iraq because I very rarely skipped a meal. Hell, it gave me something to do to break up the monotony of the day. I went back to the room and waited around until ten to start doing tattoos.

My first customer of the day wanted a cross and a banner put on his right forearm. The words "Love Conquers All" was written in the banner. Nasty was up next. He asked me to ink his last name on his left forearm. Again, that was no problem.

After Nasty's tat, I got ready to head to lunch and then on a patrol. For the past few months the majority of our patrols had originated on the Iraqi Army side of post. We loaded up on the Hummers and went there to prep for the Op Order. Going with us today on patrol was a reporter from *Newsweek*, Joe Cochran. He seemed like a nice guy and sat in as Big Red did the briefing. He took my seat in Big Red's vehicle. I rode with Tater, Donkey, and Pretty.

The patrol wasn't anything special. Chief Wahoo McDaniel handed out a lot of candy to the kids in town. We walked the streets of Balad Ruz for a little while and then remounted and patrolled from the vehicles. Because it was a hot day, I had sweated more than normal and felt like I was sitting in a swimming pool by the time we got back to base. The only thing different about today's patrol, other than having Joe with us, was that two IEDs had already been blown this morning on two of our routes. One was east and one west. That put a little more caution in the air.

Her daddy was an IA friend. Beautiful little girl.

****

From: *<Malcolm>*
To: *<Mom>*
Sent: *Saturday, August 20, 2005*

Hey Mom,

So you finally hit 100? Pretty cool (as far as for being over here). Ha ha. It was pretty dang hot over here today too. We had a foot patrol and I sweated like nobody's business. Now we know why the Iraqis are so irritable – the friggin' heat!

Things are good. I just finished doing some tattooing on my first guy from Sri Lanka. He's one of the cooks/servers working here. It was an Aztec design. Bizarre.

****

From: *<Malcolm>*
To: *<Mom>*
Sent: *Tuesday, August 23, 2005*

Just a few more days and I'll be home! Can't wait. I've been hearing all sorts of crazy tales of how well the Americans back home treat you when you come through the airport. That's encouraging. I'm glad it's not like Vietnam got to be.

Thank you for going to Shannon' memorial. That means a lot.

AUGUST 26

We left Caldwell at 0630 and headed east to clear the boundary. Before heading out, we had breakfast. We headed east on Route T1 to Jizr Naft where we cleared the bridge for traffic. That was a short operation. We dismounted, walked around and under the bridge, checked for any IED's that might have been placed over the course of the evening, and then directed traffic to resume its normal routine. We headed north on Route T3 to Route V1 then to Route D1 and Route A1, eventually doing our last leg on Route T4 before getting back on Route T1.

The route we took was interesting because I had not been on a lot of it. There was a lot of new scenery to check out. We stopped at a bombed out IA fort where we did the tourist thing and got a lot of pictures made of us hanging around different parts of the place. CFB didn't let me down on getting some good pictures. I took of few of him in front of some Iraqi propaganda paintings and on some bombed out ledges. He did the same for me.

We only saw one suspicious thing along our route that made us think someone had planted an IED. After closer inspection we saw that it was merely a universal joint used to mark a dirt road for some Iraqi. Thank God it was just that because I was the guy who walked up to check it.

We got back to Caldwell at 0930 and I headed back up to my room to catch a nap because I kept dozing off along the route. That wasn't a good thing while one was out on patrol. It put the team at risk. I started to doze off, and like clockwork, people started coming into the room asking stupid questions. CFB wanted to download the pictures we had taken. That was fine but most of the guys who came through wanted last minute tattoos.

I had closed shop because I was heading home on leave. It was tough closing up shop because the guys had been so faithful in coming to me for business, but it had also taught me that I didn't want to do tattoos full-time when I got back home. I stayed slammed with work and every once in a while I'd get that guy who would come in and be a prick. It was people like that who ruined the good times we had up in the room. On this day I finally lost my cool and put up a note that read, "The shop is closed. I'm getting ready to leave so I'm packing. Please let me. This is the nice me. If you bother me about tattoos I will be forced to show you how I can be an asshole. Peace." They still came in and I would direct them to the note on the door.

I did not think I was actually going to leave Caldwell today but one of the admin guys came by and gave me the heads up to go on leave. I was told that we would be flying out on a Chinook at 2100. Man, was I excited! I made sure that I had packed everything tight for the trip. I was good to go. It wasn't until two days later that I realized that I had left my checkbook out on my bed. I hoped nobody would mess with my account.

I had finally been able to close shop. It was a good feeling. Joe had been good to me while I was doing tattoos at Caldwell, Cobra, and Shelby. I had done a few hundred over the course of nine months. It was a great feeling walking around base and seeing my artwork on people and knowing they were happy with what I'd done and would recommend me to their friends. I looked forward to doing more when we all got home.

We had a leave meeting at 1600 and were told to meet at the coffee shop at 2000. Later, they changed the time to 1930. That was good. We got to head out earlier. We met there and then loaded on the "deuce and a halves" for the short ride over to the helo pad. The choppers showed up on time and we had a quick, uneventful flight over to Anaconda.

When we got squared away in our sleeping quarters, Atlas, Mini Me, Scottie, Harvard, and I caught a bus to the PX area for a midnight snack at Burger King and Pizza Hut. It was the first time I'd had a Hawaiian pizza in a year and a half. All I could say was it was awesome. I couldn't wait to get home.

August 27

This was our first full day at Anaconda. It was really good because, although I'd been here before, this was the first time I was able to enjoy the amenities. We went to DFAC 1 for breakfast and then trudged over to the PX area to shop for items we didn't need. After seeing the sights, we returned to the shack. The shack area for personnel in transit was pretty nice. The shacks were built in such a way that I wanted one at my house so I could turn it into a game room or some place to relax or work. They were made of wood and were on cement slabs rather than on sand covered with turf carpet. Most had good A/C and heating units. They weren't bad at all.

Around noon we headed out for lunch and then to the bazaar to pick up some sunglasses. We got the shades and then headed to the pool. It was the first time I'd been submerged in water since the summer of 2004 at Fort Sam. I can't even begin to describe how great the water felt. The Anaconda pool was amazing! It was like an oasis in the middle of this desert. It was Olympic size with three diving platforms and lane stands for racing. The personnel working there hung a volleyball net across the shallow end of the pool and three games were going simultaneously.

Obviously, the kicker for us was seeing women in bathing suits. All females had to wear one-pieces but that was still good for us. We didn't have that at Caldwell. Granted, there weren't any drop-dead gorgeous ladies but there were some attractive ones (in a combat zone kind of way). It was amazing to me how one's perception of attractiveness or beauty could change given the difference in one's circumstances or situation. We stayed for a few hours, enjoying the scenery, and then headed back to the DFAC for dinner.

That night we caught *Batman Begins* at the base theater. The theater was European. It had a balcony, tile bathrooms with group urinals, and they sold beer in the bottle. Of course it was alcohol free but it was still good. The atmosphere reminded me of the time I visited Lisbon, Portugal while I was in the Navy. A few buddies and I checked out a movie in one of the local theaters. It was the first time we had been able to drink beer in a theater, particularly, beer in a glass. After the movie we headed back to the shack for a good night's sleep.

AUGUST 28

Another day spent at Anaconda. I woke up this morning at 0700 and went to breakfast at DFAC 1. It was a nice place, huge compared to the dining facility on Caldwell. It was everything one could want in a chow hall because it was more like a real cafeteria. There were numerous food lines to ensure the fluidity of movement of soldiers and airmen. The atmosphere was nice, although it was a little quiet for its size. TVs in various positions allowed Joe to keep up with what was going on in the world.

After chow I went back to the shack. I had to be ready for a 0900 meeting we had slotted. It was one more rung in the ladder of getting out of Iraq for a while. We met and then loaded a truck to head to another location to turn in our weapons. One would think that this would have been a fluid operation since the guys who were in charge had been handling this operation for a year, but as usual with the Guard, it had its flaws. We stayed a little bit longer than any of us thought we would. Eventually we got done, went back to the shacks to grab our gear, and then were transported to a new holding area, the PAX terminal area.

Here we stayed in a tent. We picked our cots and unloaded our gear. We were told that our next meeting would be at 0700 the next morning. I lay down for a little while to enjoy being in a place where nobody was hassling me for a tattoo or something stupid like the location of my night vision goggles or if I had a CLS bag. Sometimes the little things just got so annoying.

A few of us caught a ride to the DFAC for lunch and then went to the PX area again. The PX area had Burger King, Pizza Hut, numerous vendors, tailors, and a good variety of females passing through the area on their way to and from other areas. Air Force females were the best looking females on post. They seemed to have much more positive attitudes about being in Iraq. It was refreshing.

The wait for buses back to the PAX area was long. It was good though because, again, we got to see some more good looking Air Force ladies. One could tell that they had been bombarded with come-ons the entire time they'd been in-country. I didn't want to give them another reason to hate Joe and guys in general.

When we got back to the tent area I played on my computer, sorting through my music list. Not too long into it, the power went out. In ten minutes, the tent was like a sauna. The temperatures outside were still above 100 and had been for the last four months. Most of the guys took off but I chose to stay inside and sweat out the impurities. About twenty minutes later the power came back on. It was nap time so I slept until 1800.

I headed up to the bus stop. The people there said that they'd already been waiting fifteen minutes so I walked to DFAC 2. I got there a few minutes after it closed and ran into two SF guys who were passing through and were looking for directions to the main PX. I told them where to go and which buses to catch. The time I spent talking to them was a fix. Nothing was said about their mission but my head was spinning as I thought, "Hey. That could be me. That has to be me." I left them at the bus stop and continued my walk to the other DFAC and thought about what I needed to do to become a member of the SF community. I was glad nobody else could get in my head and know what I was thinking because they would have thought I was a dork.

AUGUST 29

It was a long day of sitting around waiting for things to happen and our plane to take us out of Iraq. We were slotted to have a 0700 meeting. All that was put out was what our prospective time of departure would be. Our plane was to fly out at 2110. The Chalk Leader gave us the rest of the day to do what we wanted; our only stipulation was to be back for a roll call at 1900.

Atlas and I headed out at 1045 for chow. We caught a bus up to the DFAC and waited outside for twenty minutes for the chow hall to open. I ran into one of the Personal Security Detachment guys from Caldwell who was down to pick up our Regiment XO. He sat with us at lunch and impressed us by getting a conversation going with a cute Air Force gal.

After chow we made a quick stop at the PX for some toiletries and then headed back to the tent area. The rest of the afternoon and evening was spent sleeping and waiting for roll call and the scheduled departure. The roll call happened but the flight didn't. It got bumped back to 2310. We continued our wait.

There were a couple of good looking ladies in our tent. One was a LT. She was hot, a blonde with a nice smile. The other female had beautiful eyes and a nice smile as well. She was a bit on the skinny side but that was me just being picky. Both were very attractive, beyond combat standards. The only thing bad about them was that they were used to getting undressed by every Joe they came across. They didn't offer any play. I wondered how many times they'd been hit on while they'd been in country. I said hi to the PFC and at first she was friendly, but as the day wore on she became less and less.

Atlas told me the story of the "skate board ramp." Apparently a few guys had ordered these things prior to going home for R & R. The "ramp" was a cushion that was used to assist a couple in their sexual conquests and exploits. It lifted the female's hips into various positions and angles which allowed her to have more of a sensation during intercourse and allowed the male deeper penetration. The thing that was so funny about his story was that he had been telling his wife that he had one on the way for them to use when he got home. In reality, he hadn't ordered it but we were sure she wouldn't have minded.

August 30

The day didn't seem like it ever began; it was simply a continuous run from yesterday. We spent the entire evening waiting for our flight. First we were told it would leave at 2110 last night. Then it was bumped to 2310. Next, it was bumped to 0030 and finally it was bumped to 0130. We boarded the C130 like a mob of zombies. Our only relief was that we had finally caught a flight out of Iraq.

The flight itself was uneventful. There were a few dips along the flight path because the load master later told us that planes had been taking fire from those areas in the past. He said they were dropping flares or something to that effect. I didn't quite understand it but I guessed it was some kind of countermeasure. Fortunately for the aircrew, most of the ground fire was from AKs.

We arrived at Ali Asalim at some point in the wee morning hours and were transferred from bus to bus until we got to an area where we turned in our IBA and Kevlars. We were herded around like cattle. All we wanted was some hot chow, a shower, and a good night's sleep. Getting rid of the armor wasn't a bad gig. Without the weight of those things a man was able to walk a little straighter.

After turning in our battle rattle we were corralled into another holding area where we showed our IDs and were given billeting. The guys I was with got one of the tents farthest away from everything. That bummed us out until we got to the tent. It was actually pretty nice, cool, and clean. We finally got chow at 0630. The DFAC at Ali wasn't bad at all. It was small but it served good food and the lines moved quickly.

After chow we headed to another formation for the handout of flight tickets. Next we went to a meeting that told us of another meeting happening at 1300. That seemed to be the natural state of the military in general but the Army more specifically. Joe went from formation to formation, meeting to meeting, all in the name of accountability. So during the day we took showers and slept and then went to our meetings.

The 1300 meeting was the big daddy. In this meeting we were told about time lines, movements, and customs. After the briefs, we weighed in and then headed to the customs tent to let them check us out for anything non-allowable in to the US. It went relatively smoothly. The Navy was in charge of this particular portion of out-processing. They checked everything in our bags, scanned us, X-rayed our bags, and gave us an amnesty period to get rid of things before the whole customs process began.

After clearing customs, we were shuffled into a holding area called "Freedom Tent." The tent was just a big hanger looking thing with chairs and a TV for entertainment, not much else. Rumors were already going that our flight was going to be delayed.

The Path Less Taken

From: <Mom>
To: <Distribution List>
Sent: Thursday, September 01, 2005

Malcolm got to Nashville yesterday at 4:20, an hour earlier than first scheduled. He looks GREAT and seems just like he was before he left! We drove home and he kept saying how beautiful and green Tennessee is!! We even saw some deer on the side of the road. We asked if he wanted to go home for supper or to a restaurant and he picked the Cookeville Outback. The folks there were so nice and kept coming up and thanking him for serving over there. The manager kept checking on us too and paid for Malcolm's part of the meal. We saw several friends there so it was like old home week with lots of hugs from guys and gals.

I think he likes the welcome home sign and the yellow ribbons in our yard. And Eliseo and I had matching tee shirts with welcome home Malcolm on them. We got to go right to the gate to meet him at the airport and when he and a Marine came out the door together several waiting passengers started clapping. The Marine with him is from nearby Gordonsville and we gave him hugs too. His parents had not gotten to the airport yet as they might not have known his plane was early. But he told us to go on and he'd be all right.

This morning we got up at 5:30 and he was already awake. We had some breakfast and then took an early morning walk to the golf course – all so green and misty. Saw four deer and a rabbit there too. A great time was had by all and we are just so thankful to have him home. He'll be here until the 16th.

You may not hear from me in awhile.
He has a lot of things he wants to do, not just sleeping!

****

From: <Traci>
To: <Mom>
Sent: Thursday, September 01, 2005

Margie,

Wow!!!!!!! Thanks for sharing your experience with me!!!!! I must admit I got teary eyed; I could just picture reunion in my mind! I'm so happy for you … there is just a sense of overwhelming peace being in your soldier's presence and I can feel that coming from your words. Go enjoy your son and the time you have with him! (And your son had GREAT taste … Outback is the best place in town!!!!!!!!!)

****

From: <Colonel Eck>
To: <Mom>
Sent: Thursday, September 01, 2005

GREAT NEWS! WE ARE HAPPY FOR ALL OF YOU. BE PROUD OF YOUR SON, WE ARE. THANKS FOR DOING SUCH A GREAT JOB MALCOLM. I KNOW THAT IT WASN'T EASY OR PRETTY. WAR NEVER IS AND NEVER WILL BE. ALL THE BEST. COL. ECK & JUNE

SEPTEMBER 9

I've been back in the States on leave for the last nine days. It's been sweet. We flew out of Kuwait on the 31st and made our way back to the States with stops in Hungary, Ireland, and Canada before landing in Dallas. When we hit the Shannon, Ireland airport, more than a few of us bellied up to the bar for a few quick drafts. They tasted great, a welcome reminder of the good things in life and why we were fighting to keep them good. Flight time was about 20 hours. That was the worst part of the trip. Too long in the air in little seats with big guys sitting next to us and bad head phones (for listening to the in-flight movies) was not my idea of a good time but we made do.

As expected, when we touched down in Dallas, a round of applause filled the cabin of the plane. Joes were very excited to be back on home turf. It had almost been a full year since I had been in the States. Over the past 18 months, I had only spent a total of 20 days in Tennessee. I was ready to be back in the US even though I had great offers to visit England and Amsterdam. We departed the plane and headed through another quick customs check and then were directed out into the main terminal area. As we entered the area, about 40 or so people greeted us with signs, a flag, cheers, hugs, and welcome home gifts. It was very nice. My hat went off to those volunteers who repeatedly showed up to welcome the military home from abroad. I had the good fortune of shaking hands with a gentleman who had served as a Platoon Leader with the Big Red One during the WWII. He was very proud that I was wearing the combat patch of the 1st ID.

I flew from Dallas to Nashville with a young Marine who was from Gordonsville. He was a stand up guy. He said he worked with satellite information, comms, and other technologically advanced equipment. I told him not to worry about it, he was still a Marine and he and his guys were still doing a hell of a good job kicking ass in Iraq. We had a good laugh. When we got off the plane in Nashville, Mom and Dad were waiting at the gate. A small applause went up for us as we reunited. Mom gave the Marine a hug too as his parents and friends hadn't been at the gate to welcome him off the plane. He said they were still on their way after he made a call from Dad's phone. We shook hands and then departed.

<u>September 11</u>

September 11th. It had been four years since those assholes attacked America. Four years ago that one chain of events did more to change the direction of my life than anything else had up to that point. I felt called to join in the fight against terrorism. I made decisions and left behind things that were very comfortable and good to go to a war in a country that I could have cared less about. We were "rebuilding" a nation that I still didn't care much about up to that point. It was great to be home.

Yesterday was a good day. I got up early and started cleaning my room. Mom cooked me breakfast. Dad had already left the house to go work at the Presbyterian Church's book sale. When I finished cleaning the room, I hit the shower and got ready to head into town to enjoy the city's Fall Fun Fest. The event was a celebration of good music and tasty barbecue. The city had hosted it for years. It was an ideal time to see some folks I hadn't seen in a long time. The weather was beautiful that day.

I pulled into a back parking lot near the church and then headed over to the tent where Dad and Mom were selling books. Some of our family's old friends were there and I answered the typical questions and asked the same ones I'd been asking since I got home: How are you? What have you been up to? How's your family? Any crazy things happen in your life while I've been gone? The questions were lame and more often than not, I already knew the answers they were going to give me. It was just a matter of going through the motions so people could see that I'd come home in one piece … at least up to that point.

Dad and I decided to walk the streets in search of some lunch. We came into the festival proper from the northwest corner where the kids were being entertained. It was a good set-up. A skate park had been erected for them to use. Granted, it wasn't anything like Tony Hawk would have designed, but at least they had something in the middle of the festival that would be entertaining to them. Many of the skateboarders in Cookeville got run off from places around the city. All they wanted to do was have fun. I'm glad someone finally got some cahoonas to do something like that for those kids.

I ran into quite a few friends along the way. The bad part about being away and then coming home and seeing a bunch of friends in a localized place was that whomever I was with had to wait while I talked to everybody else. It was funny and sometimes overwhelming because I wanted to talk to everyone I saw and give them my attention, but most of the time it ended up being like a sound byte on some media channel. I only got in the superficial pleasantries and then moved on to the next person.

I ran into another medic who'd been with us in Iraq. He'd been sent home on a hardship discharge. It was good seeing him but I wondered how he was doing with his situation. His family was with him so I didn't ask him too many questions. I gave him a hug, made some small talk, and then moved on down the line.

I had also taken the prescription for my glasses to get some new ones made. The glasses the military issued, BCGs (birth control glasses because you'll never score with them on), helped me out quite a bit when I was typing on the computer so I thought I would get some more attractive looking ones for the long haul. I didn't realize how expensive glasses had gotten over the years but they were a necessary evil in my eyes so I coughed up the $200.

After leaving the eyeglass store I headed to Nashville for a reunion with some of my family members, cousins whom I hadn't seen since the last member of our family had died. It was sad how life so often got in the way of everything one wanted to do while living. It would always be like that but it still sucked. There was so much that our extended family had to offer; and unless we made the effort to be an active part of each other's lives, we would lose out on something special in our own lives.

The biggest reason I had for wanting to go to the reunion was to see my Uncle Clarence. He was a warrior from a long time past, when warriors were proud and there was still a romantic lure to being in the service of the nation. He'd seen the bad side of mankind as a prisoner of war. He'd also seen the good side of mankind as nations forgave each other after those wars and helped one another to rebuild what had been destroyed. After his military service, he was a lawyer in Nashville for years. He was one in the handful of men whom I've met in my life whom I respected simply because he was the epitome of a true gentleman. He was a humble and sincere man.

When I was growing up I had some bouts of rebellion. Okay, so I still did to a degree but what I remembered most about Uncle Clarence was that even though he might not have agreed with what I was doing at the time, he never made me feel like he was judging me. And when I served first in the Navy and now with the Guard, he was proud. He welcomed me into the brotherhood of warriors who had given the best parts of their lives, and in some cases, their lives, to make this nation great and safe.

I have always admired him. Yesterday, we sat for a while on his back porch, just he, his son Winston, and I, talking about what was going on in Iraq, things he had experienced in his time in the service and about my cousins Mike and Giles who were also serving. Mike, Giles, Clarence, and my grandfather were all West Pointers. Tradition ran strong through their veins. Although I'd only once visited the United States Military Academy, I took pride in knowing that some of the finest leaders in the Army have been and were related to me.

My night was already planned so I did not get to spend a lot of time at the reunion, but the time I was able to spend there was so touching. As I said my goodbyes, tears filled Uncle Clarence's eyes. He'd been there; hell, he had been a POW. He knew what I was going back to and he was proud that I was still committing to do it. It took everything I had in me not to break down. As I wrote this I was fighting back tears. Honor and pride swelled within me. I was the fortunate one because yesterday I was able to share that moment with him. It was very humbling.

Winston walked me to my car and he looked a little shaken as well. It was a bitter sweet departure. Winston's son Giles was getting ready to move into the active work of the Army. Winston understood what that meant. He'd heard about war from his dad, uncles, and now cousins. Through the journal entries I've sent, a picture was painted about which I'm sure Winston was concerned. He knew Giles chose the path of the warrior and he supported him 100%, but Giles was still his son and as a dad, he worried. Thank you, mothers and fathers, for caring, loving, and supporting your warriors. I hope to God that none of you have to experience the loss of one of your own; but I know that you know the possibility of it as much as we do. God bless you for your strength.

The evening was much lighter. Two of my classmates from high school had planned a trip for us to see Motley Crue while I was home. I met them at a local Mexican restaurant, just down the road from where Uncle Clarence lived, and we headed to the old Starwood Amphitheater for an evening with one of rock and roll's most decadent bands. It was a blast! We were supposed to hook up with some friends of theirs from Nashville. We found them among the people in the grass section of the venue. I had the incredible fortune of running into the brother of the first girl whom I had dated in high school. He was with one of my former Young Life gals. They were dating and he had surprised her with a ticket to the show.

Brando has always been a great friend. Every time I've run into him, he's always been so positive and full of life. This night was no different and even more so because he was with Lizard. I hadn't seen Lizard in years. She was a beautiful gal in high school but she had blossomed even more so in those few years since I'd seen her last. We talked for quite a bit before the concert started and I got caught up on how life was treating her. She was in her second year of pharmacy school at Stanford. There wasn't anything better than a hot drug pusher at a Crue concert.

The show was incredible! The original band members Vince Neil, Nikki Sixx, Tommy Lee, and Mick Mars lit up the stage. Tommy, always being the one to take things to the limit, broke out the "titty cam" and was successful at having numerous ladies on the front row flash him for everyone else in the venue to see on the big screens above the stage. They belted out all of their old hits. It was a blast seeing and hearing them onstage, together again, and being there with people with whom I enjoyed hanging out.

There were some rumors that Mick Mars was not going to be able to get on stage and jam that night because of the degenerative bone disease that he's struggled with his entire life, but he was there and he kicked Nashville's ass! This may sound weird but I felt like I was standing in the presence of another warrior. That man had to be in so much pain while he was up on stage. He was hunched over the entire time because of the disease yet he hit every note and rocked the house like nothing was bothering him. My hat went off to him for pushing through and giving us another great performance. None of us knew if he'd be able to do the next tour.

After the show we went to Jonathan's, a local tavern, for some food and drinks. Menus were passed around and I ordered a Hawaiian pizza and a beer. Lady luck was with me as I was seated next to two of the most gorgeous gals I'd seen since I'd been home, Tab and Nix. They seemed really cool and appeared to have their acts together. So much about them was attractive to me besides their physical beauty. Granted, they didn't give me the time of day, but they sure were nice and nice to look at.

****

From: <Trevor>
To: <Malcolm>
Sent: Monday, September 12, 2005

Malcolm,

Have you ever thought about being a writer? For obvious reasons, I was in tears when I finished reading this. Then Stephanee came in the room and asked what was the matter and I showed her your letter and then she broke down. Many things have been written about my dad but nothing has ever summed him up so well and for that matter the spirit of the clan. Yeah, he was tough to have as a father. I think I started boot camp before I was out of the cradle but then again I was too young to remember. And yeah I was rebellious too and still [am] somewhat today. And we had our differences. But when all the dust settles, you have to admire, respect and love him. Glad the night time gave you some R & R.

****

From: <Malcolm>
To: <Distribution List>
Sent: Friday, September 16, 2005

Hey folks. Just wanted to let you know how much I appreciated the warmth and love I felt on my return trip home. It was awesome! I thank you all for giving me your best whether I got to see you in person or on the phone or though an email. You make me more proud to serve the people of this great nation.

There is so much more to write about leave but I won't bore you with it. Just wanted you all to know that I love you and can't wait to be back home for good. It's going to be one hell of a road trip!

If you're in the Cookeville area and you get a chance, listen to the Country Giant at 0840 today. I'm getting interviewed by the best DJ in Tennessee, Gator Harrison. If you're not in the area he said you should be able to log onto the website and hear it over the Internet. Gator and the Country Giant have bent over backwards to support the 278th through our entire deployment. My hat goes off to him. And a hearty "CONGRATS" to him and his wife on the upcoming birth of their first child.

I got to spend my last night in town with friends and family. My cousin Winston was passing through town on his way to Knoxville to try a case and we had the good fortune of breaking bread with him. He told us some great stories of his dad, my Uncle, Clarence. He was even more [of an] amazing man than I had known before and one of the great opportunities he had in the past year or so was to see his grandson, Giles, graduate Airborne school. And to add icing on the cake, Fort Benning made him the guest of honor! Winston said it was an incredible experience and I can't wait to see the video from it.

I know God is watching over us and I know you all will continue to pray for and support us. Thank you. We never could have done what we do without you folks back home. God bless you all.

September 16

One of the things I had wanted to do while I was home was visit a chiropractor. After a year of wearing body armor, gear, ammo, and riding around in military vehicles my back needed to be serviced. Ken Schmitt-Matzen was gracious enough to allow me to visit his office almost every day while I was home at no expense. Again, my thanks to my fellow Americans who know how to treat their soldiers!

What a day! We headed out to Nashville at 0745 so that I could catch a 1015 flight to Dallas. On the way we listened to the Country Giant radio station. We were waiting for the 0840 playing of an interview that I had done with Gator Harrison the day before. Gator has been one of Tennessee's top radio personalities for years. I've known him for some time. He and the Country Giant offered any 278th soldier the opportunity to be interviewed while home on leave. I decided to take them up on it because they had been so good in relaying messages to family members while we were in Camp Shelby and in Iraq. From what I could tell, they were our number one fans outside of our immediate families. What they did for us was amazing. Thank you.

I met Gator at the station on the 15th and we chatted a bit about what was going on in both of our lives before he offered me a spot on the radio show for the next morning. We had a good time doing the interview. He, Stickman, and Big Stu were their regular radio personalities and made me feel totally comfortable on the mike. Gator asked how things had been going in Iraq and I shared my thoughts and feeling about some issues and told the story of the insurgent's life we saved in February.

When he played the spot this morning, he also played the song "Homecoming" that I had written while I was at home. It was exciting to say the least. One of my life long dreams had come true. Mom and Dad were proud as they cranked up the volume in the car. It made me laugh.

HOMECOMING

Why won't you let me come home?
I have been gone for too long
I've seen some things I hope to God
You'll never see but that's just me

Seems like forever and a day
When I kissed your cheek then walked away
I've been some places that I know
Were not too good but that's understood

I miss you so much that it hurts me
When I think about you I almost cry
And you know I chose to be a soldier
But don't ask me why

But I'll be home this year for Christmas
Might even catch Thanksgiving too
Tell all my friends how much I love them every one
And I'll be coming home real soon

We've spent a year out in the desert
Seems like forever in the sand
We've seen some good times
And we've had some good friends die in that barren land

But you know our time there's almost over
That makes the days there fly on by
I'd do it all again to keep this country free
Now, don't ask me why

But I'll be home this year for Christmas
Might even catch Thanksgiving too
Tell all my friends how much I love them every one
And I'll be coming home real soon

I'm just a simple guy
With simple dreams of the simple life
But I'll stand up when they call for me
Cause the life we live is worth living free

****

We got to the airport with no incident although we did see one wreck on the way. A nice Beamer had been tail-ended by a tractor-trailer. It didn't look like anyone was injured but the car sure was a mess. As we rolled up the off ramp to turn on the road that would take us to the airport, a stalled car sat in the middle of the road. Immediately, my alert status went to red. If we had been patrolling in Iraq we would have halted the convoy and ensured that the vehicle in the road was not set up as and IED. If it was a VBIED we would have cordoned it off, stood off a few hundred meters, and waited for EOD to venture out and blow it. As it was, we were in America and fortunately, we could still feel relatively safe on the streets.

I flew Delta Airlines out of Nashville. I wish I could remember the name of the ticket confirmation lady. She rushed me to the head of the baggage check-in line and confirmed my seating. She also gave me first class! It was awesome! I'd heard that some of the airline personnel were doing that for soldiers but it hadn't happened to me until that time. I'd also heard that some of the folks who traveled first class would give up their seats to a trooper in uniform. Thank you, America.

My first flight took me to Dallas. I arrived a little after noon and hung around the USO, using the computers that they had available for us. While checking my emails, I came across one from a gal I hadn't talked to in months, Sarah. She had been in Colorado for the last year and had fallen in love with a guy she had worked with at Wilderness, a Young Life adventure camp. I was really happy for her although I'd never met the guy and had no clue as to what he was like. I was certain he was quality though. She had a huge heart but I had never known her fall in love too easily. I was sure she prayed about their relationship fervently. I called her while I ate my last

American meal and enjoyed hearing her voice, a voice filled with joy, excitement, and expectation. She sounded great. I was excited for her, her love, and her future.

Our plane was slotted to leave at 1500 bound for Kuwait. I didn't know if we'd take the same route as when we flew home but my bets were on it. We met at the manifest area and received our assigned boarding times. We were told that we were not actually going to leave the States until 1830. A few of us decided to hang around the airport restaurant / bar. We ate and had a few beers prior to boarding. I got to know a few soldiers from the Montana National Guard. They were part of an air wing stationed at Anaconda. They had left Iraq with us but we hadn't made any contact with them on the way out. Now, we all hung out together like a herd of rabid camels. Amanda became our designated sober representative and she made sure we all made it to the plane on time.

We had the good fortune of meeting a fire fighter from Los Angeles named Mike. He had been working the Katrina disaster in Gulfport, MS for a few weeks and then had been called back home for the birth of his son. All of us felt proud to be in the same room with him because we all wanted to be doing what he had been doing: helping our American brothers and sisters back home. We were all Guard and that's what the Guard did. We helped during emergencies and natural disasters as well as in combat. He showed us some pictures of the devastation he'd witnessed and said that his pictures could not do justice to the wreckage that was once Gulfport. Mike had volunteered with FEMA and said he hadn't done any fire fighting work but more work along the lines of public affairs. He was one of the guys to whom people directed their complaints. He was their shoulder to cry on or their hand to shake in saying thank you for the help. "Incredible" was the best word he could use to describe his experience.

The flight back to Kuwait was even less eventful than the one on the way home as one might imagine. We were going back to war – excitement was nonexistent. We only stopped in Budapest this time but were not given the opportunity to disembark. It was lame. Two really good things did go down on the trip. One was that I got to ride back beside Amanda. She was fun and in a timeless kind of way, appeared more attractive than she actually was. We shared some stories and she told me a lot of good things about Montana. Of course, I had to hear stories about her boyfriend, but that went with the territory. It was a good time hanging with her nonetheless.

The other thing that was cool was when I went to the front of the plane to relieve myself. It wasn't so much relieving myself as it was that while I was in line, looking back through the cabin, and I noticed a Captain directly in front of me. He had been my roommate while we were in seminary! I gave him a hug and we caught up until the lavatory was empty. When I was done, I suggested we hook up when we touched down in Kuwait.

My old roommate in Seminary. We hooked back up inflight to Kuwait.

Upon arriving in Kuwait we boarded buses for the hour-long trip to base camp. It too was uneventful but made me wonder why no terrorist had figured out that our routines in movement there were very predictable and totally unsecured. All the soldiers and Marines filed into a large tent to be given a safety brief and instructions as to what would be going on for the next 24 hours. Cliff and I sat together and made fun of the whole process. We were then escorted to the area where we would retrieve our body armor and helmets. It was depressing because once we got our armor we knew we were back in the game. Happy days were gone again. After receiving our armor, we were taken to the billeting tent where we were assigned to our respective tents for the evening.

****

From: *<Gail>*
To: *<Malcolm>*
Sent: *Monday, September 19, 2005*

Hi,

I will be glad when you guys come home. I just think it's way past due. We do support & love you guys so very much. Indeed, you are our heroes!! The first grade class sent you the pictures. We will send some more. Please remember, this is 1st grade, so their writing & coloring, well, we are practicing. I just let them write & do what they wanted. Your mom has been great to us. We do appreciate everything that has been done. Brad [a Marine], our son, was killed last year in Iraq, so please take care of yourself & take no unnecessary risks, just come home safely.

SEPTEMBER 21

I got back to Caldwell yesterday at 1930. We missed dinner chow but not to worry. I came back to a room full of boxes that had been sent by the great supporters back home. There was food for days in them. Dinner was beef jerky and Snack Wells. It was good stuff. I also received eight boxes of paperback books from my cousin Winston's wife. We had various libraries in random buildings around base. Most were sent to Joes by friends and family but some were sent through the USO and other organizations that supported troops.

A 386th convoy had arrived two days ago and we saw some of the guys at the PX. We asked them to tell their people that they hadn't seen us so that we might stay another night or two before going back to Caldwell. The next day we were picked up by another group of 1st Squadron guys. Our convoy left Anaconda at 1600. We stopped once at FOB Warhorse so that EOD could do some stuff with Warhorse's EOD. It wasn't a big deal and the stop gave us the opportunity to stretch our legs and check out their PX. While there, we stocked up on Gatorade and magazines for the remainder of the ride. We rode in the back of a five ton with armor plates welded onto the bed walls for our protection. We could barely see over the metal sheets, so as far as scanning the area as we went, it was pretty much futile.

Our time at LSA Anaconda was too short this time. We only had one full day there but we made the most of it. A few of us left the transition area around 1030 and headed to the pool to hang out for most of the day. At 1500 we planned to see *The Fantastic Four* at the base theater and then stay for two other movies later that evening.

As I said before, we ran into some of the 386th Engineers while we were in the PX and they said that they had room on their convoy for some folks if we wanted to go. We quickly declined, letting them know about our plans for the afternoon. They didn't blame us and said they wouldn't say a word if anyone asked.

The water felt great again. We were ecstatic because there were actually some good looking military girls at the pool. That made things easy on the eyes. Most of the girls who were at the pool came in with a herd of guys or one or two of their girlfriends. Unless you already knew them, you weren't getting close.

The movie wasn't bad either. After the first flick I headed back to the transition area for a combat nap before the 2100 showing of *The Island*. I had bought a pirated copy of the movie while at Caldwell and had liked it but something about it made me feel like some of it had been left out. Besides that, the audio quality was awful on my copy. I was right. The theater version was much better and had quite a bit more to it than the pirated copy. After the movie we headed back for what we hoped would be a good night's rest. It was still hard to get back on a normal routine.

Now that I was back I regretted ever having left home. Caldwell still smelled like a bag of smashed ass and once again, it was browns and tans everywhere. I already missed seeing the deep greens of the Tennessee landscape in summer. Fortunately, the ordeal of being back in Iraq was only for about 30 more days and if I could get my attitude right in the next 24 hours, it would be all right. That's what it was all about anyway, attitude. Sure, I could have whined and bitched about having to come back, especially with the situation still going on in the Gulf area, but I just couldn't afford to do it now with so little time left. I had to get my head right so that the next 30 days wouldn't cost me or anyone else I worked with our lives. That would be unbearable this close to coming home.

I'd gotten the word that we still had a very active schedule from here on out. That was good. It would keep us on our toes and help us to pass the time as quickly as possible. I hoped and prayed that all of us would get our attitudes right for the last leg of this thing. We couldn't afford to be complacent or think too much about going home. We had to finish a job and that was our priority at that moment. Sure, we'd think about home every waking minute but we couldn't let those thoughts or the emotions that came with them supersede what we had to do.

The light at the end of the tunnel was real and it was shining brightly. Soon enough we'd be home and soon enough we'd be back to the people and things we missed so dearly. I couldn't wait! I didn't get to see or do enough while I was home. I was ready to get back and suck the juice out of life.

September 24

Life's been pretty slow since my return. I've only been outside the wire one time. Two nights ago I awoke at 0215 for a 0300 mission on Route C1 and Route V1. Our mission was route clearance of Route C1 and then do an over watch on Route V1 to make sure no jackasses were placing IEDs. I drove Dark Wing's vehicle. It was like old times. It wasn't a bad gig other than we were acting as minesweepers. Our whole purpose was to make sure that the first unit coming through on convoy in the morning didn't get hit by an IED. We put ourselves at risk of being maimed or killed to ensure that the risk to the personnel in the convoy was lessened.

Fortunately on this night, there was no action. We did see quite a few dogs along the side of the road and I tried to run a few down along our way. Those dogs were quick when they saw headlights aimed at them. The routes we had to use in our AO were horrible. Though there weren't many explosion holes on Route T1 or C1, the streets were rough from 150 degree weather and heavy vehicles running over them for years. I'm sure some of it was from wash out as well even though we hadn't seen rain in months.

At one point in our mission we entered another battalion's AO. We came upon their checkpoint as they were fortifying it. It was quite impressive with its ten feet tall barriers as well a score of soldiers. They allowed us to pass and we continued our mission to the next checkpoint which was the end of our AO. After turning around, we headed back up the road driving in blackout condition. We did this for a few miles then pulled off the road to set up an over watch position.

It turned out to be a funny situation. Pretty, Tater, and Crowfoot were in the vehicle behind us. Pretty and Crowfoot had been in Scout Sniper platoons when they were on active duty with the Marines. They were our tactical geniuses ... or so we thought. They only pulled off the road a few feet initially. Dark Wing had to get on them to move and when he did, they set up their vehicle in such a way that if anyone driving the route had come by with his lights on, he was sure to see a reflection from Pretty's vehicle. Dark Wing was beside himself as to their tactical inefficiency. He slammed them on the radio and we all laughed. Our laughter hurt Tater's pride and he started talking smack back to us. That's how we passed the time when we were slotted with lame missions. We talked smack to one another over the radios; so much for noise discipline.

An over watch position was set up to observe a certain area, in this case the road, for IED activity. Our unit would be in blackout, using NVGs, and would engage any person who looked suspicious or was obviously placing IEDs. That night we had no activity and after a short while we headed back down our original route. Usually in Iraq we drove the center of the road and at a high rate of speed. With NVGs we could still do that but it was much more dangerous. Dark Wing asked me to slow down and cruise the road so we could observe a bit more carefully what was going on beside the road.

After reaching the other battalion's check point, we turned our lights back on and continued our patrol in normal mode. We returned to Caldwell as the sun was coming up and breakfast was being served. The rest of the morning was spent in the rack, catching up on sleep. The afternoon was spent in the gym and cleaning up, preparing to come home.

SEPTEMBER 29

Tonight was the last fight night. It was entertaining to say the least but not really one of the better ones. A few of the guys bowed out early in their matches, I guess due to exhaustion. It was a bummer because we all got so fired up to see good fights. I'd say that the best fight was from a little guy named Bo. He wasn't the one we had picked to win but he did. He fought a LT who couldn't hang with him. It was a good victory for Bo.

We did go on a mission yesterday to take pictures of a shrine that had been blown. Our unit has been to this place numerous times and every time we've come back saying, "Why in the hell did we go down there?" This time was no different other than we got to see the destruction that man can do to the things he creates.

Shrine destroyed by militants. That's Donkey on the front right.

The mosque had been a large building that appeared to be relatively new. All that was left was the front wall and an out building. Debris was spread out over the area for a few hundred meters. Bird feathers were mixed in with the debris as some of them were caught in the blast. No humans appeared to have been in the building when the bomb exploded; at least there was no odor from human decay in the debris. We couldn't figure out if it was a bomb placed on the inside the building or some kind of vehicle-borne IED that was driven into the side and detonated upon contact. Whatever it was, it leveled the place.

When I walked into the wreckage all I could think about was the old heavy metal promotional pictures I had seen growing up where the bands were standing amidst some old, torn down building. It was the same here, only now there were soldiers with weapons walking amidst the ruin. We took a few pictures in and of the rubble and then a few of us scouted the surrounding area for other carnage. The explosion threw some things a few hundred meters away. We found pieces of metal that were similar to the metal from a vehicle. They were riddled with shrapnel holes. Whatever had been used was effective.

Fortunately, as I said before, the only bodies we found any evidence of were of pigeons that hadn't expected the blast. They were everywhere. I would like to have seen the actual blast that took the structure down. It must have been big. Those birds did not have a clue what hit them.

Our new Platoon Leader was with us on this trip. Rugger had come to us because Big Red had been bushwhacked by severe back problems. He'd had back surgery in the past and something went wrong while he was in Iraq. I suspected it was from wearing the IBA. He was sent home. Big Red had done an outstanding job as our Platoon Leader. He would be sorely missed but I was glad that he was getting care for his back.

Rugger was one of the casualties from the April 4 ambush south of our battle position. He suffered a gunshot wound to the left shoulder, deltoid muscle. After a few months of rehabilitation, he volunteered to come back to our unit. At 24, he'd already been in firefights, been wounded, and had led two platoons of combat tested men. He was different but he was all right.

Of course with new leadership there was always the agony of suffering through change. It was tough for Dark Wing because he did not want anyone else to have control over his men. Rugger wasn't trying to outdo him or take away the control he had of his platoon, but I thought Dark Wing felt his authority was somewhat threatened. That was a fault of his that still hadn't settled with me. He always wanted to degrade anyone higher than he in the Chain of Command in front of his guys. I guess he felt it made him look good to some of the other men in our platoon when he did that, but to me he just came across as insecure.

To a point, I could agree with him on the leadership we'd seen in the 1st Squadron of the 278th and at the Regiment level. Leadership had been very disheartening in many ways. The officers had a tendency to micro-manage. More than a few of the senior NCOs in administrative positions were only looking out for themselves. They would never go above and beyond their station to assist the men whom they were charged to help. That was their job and they would not do it. It seemed like the only time that they would move on something was if it reflected positively on them in the eyes of their superiors. The rumor, that ran rampant everywhere on Caldwell, was that a large number of the people in administrative positions, enlisted and officers, were putting themselves in for Bronze Stars and Army Commendation Awards simply because they were in a position to push through the paperwork. Issues like that made the administration look self-serving. They made something honorable, the award process, leave a bad taste in everyone's mouth, especially the guys who'd dodged bullets and weren't getting any award for their competency in combat.

I believed whole-heartedly in giving awards to those soldiers who had earned them. If one did his job and did it well, by all means, he should receive an award because that was probably the only "thank you" he was ever going to get in the military. But we had folks who had gone above and beyond the call of duty and weren't getting anything because the people who pushed the paperwork were jealous of the awards. Somebody ticked somebody else off so the paperwork wasn't going to be sent through the proper channels or given the proper respect that it should be given so that soldier could receive the award he or she was qualified for and had earned.

The most disturbing issue was that so many of the awards that were submitted were submitted by someone putting himself in for the award. I seemed to recall a certain Presidential candidate who had done that in Vietnam and I didn't respect that he'd done that when I had heard about it. Granted, a lot of guys would be overlooked, but I was willing to bet that none of them would submit the paperwork on himself if he had to do it just to get a ribbon or badge. They knew what they did. Self-promotion lessened the value of any award earned by someone in the theater.

I tried to prove my point by continually asking Dark Wing and Big Red to put me in for the Combat Driver Award. I was in a qualified slot and had driven the miles in country for the award but I didn't get it. If I had I would have only worn that badge when I was in Class As. It was a joke to me.

We had a ceremony to distribute ARCOMMs. We'd see which one put himself in for more awards than deserved. Big Boss Man, who had already served 28 years, stated it the best. He said, "The only award I want is waiting at home for me." I was in agreement with him. Just getting home safely would be the best award. No other award could match that.

Turn the Page

October 1

I woke up today at 0500. Our mission was slotted for 0800 but I wanted to get up early and get some chow before heading to the track line to get on the Bradley. I'd almost gotten to the point where I'd forgotten how much I hated riding in the Bradleys because we hadn't used them in so long. But very quickly, the reason came back to me as to why I did hate them. They were loud and cramped like a hooker in the middle of a scrum.

We mustered at 0700 and waited as everyone else showed up for the Op Order. Things were very light hearted as we did our last minute checks. I didn't have an aid bag with me so I borrowed one of the Combat Life Saver bags that CFB had with him. It would serve the purpose for our operation today.

Our new CO did the briefing at 0730. Black Water 6, like Big Red, had been sent home for back surgery. The Captain did a quick, concise brief. I appreciated that he didn't treat us like a bunch of green-backs. We had been in country for almost a year and had done hundreds of missions. Shrek led us in a quick prayer before we all loaded onto our respective vehicles and awaited the word to move out.

Riding in the back of a Bradley was like riding in a cast iron tomb on wheels. It sucked. I hated everything about it. The only good thing about it was the fact that it did offer us a little more protection on the sides than our counterparts had in the 1114 Hummers. We rumbled the 14 miles into Balad Ruz with no incident.

Our mission this day was to go into a southern sector of town that was predominately chop shop and auto repair work. We were to search the area for any signs of bomb making material or vehicles set up to carry explosive devices. We'd been there numerous times before but as the elections drew near, the CO wanted us to keep the people at bay by our presence in the area on a routine basis. Our entire company went. Tre-Deuce's job was to search three adjoining streets and the shops that lined them.

We dismounted and set security first thing. After determining that there was no immediate threat, Dark Wing gave the order to move out and I took point for our first route. He made the comment that he liked having a medic walking point. It made me laugh too, considering that wasn't the best place to have a medic if something happened and the point man got waxed.

Since it was after 0800 when we arrived, the businesses were already open and conducting routine work. We walked the first street, made sure it was secure, doubled back to conduct searches of the buildings, and then handed the reigns over to Red Platoon so that they could take over the intersection. We started targeting shops, looking for anything that might be suspicious. Some were brake shops while others were engine repair. A few were paint shops. All in all, it was a normal day in town. Nothing out of the ordinary stood out enough for us to bust anybody and tear their shop apart. There was so much crap lying around we would have missed bomb making material anyway if they had really wanted to hide it.

In a couple of the shops I did find cell phones that had been put in places where the normal customer could not see them. In most countries that wouldn't be a big deal, but in Iraq we had to treat the incident as if it could be a volatile situation. We took the time to ask questions about the intended use of the cell phones. Cell phones had been used to detonate explosive devices. Fortunately for us, none of the phones that I found set off any IEDs while we were pressing buttons. That wasn't to say that they would not be used at another time, but at least for that day, they weren't being used to kill Americans or Iraqis.

We spent about three hours in town going through streets and shops. Eventually, the CO called the mission and we returned to the

Bradleys and then to base. I ran in and took a shower to get the dirt and sweat off me and then headed to chow. Normally we would have to wear our DCUs to lunch, but because it was Cavalry Day, we could wear our physical training uniforms. It was nice too because it actually wasn't too hot outside. It was in the high 90's or low 100's.

October 3

I've been trying to keep myself busy doing things to pass the time since I don't have my tattoo equipment in Iraq any more. Doing tats was great. It always gave me something to do when I had down time. Since taking the stuff home I've had to come up with things to do so I wouldn't get bored or into trouble. Most of the day has been spent between the gym and the chow hall. Dark Wing and I got up early for breakfast and then I went to the gym at 0900 because I was going nuts just waiting around for the "1000 work out time." I played Emily in a few games of ping pong and beat her. Usually I played her after I worked out and she often won. Playing her before my workout kept her in check so that she wouldn't think she was all that good.

The mission we pulled yesterday morning was a joke. We were told to be at the vehicles by 0030 but once there, we were told that our part of the mission wasn't going to start until about 0300. We could leave base at 0300, arrive at our checkpoint, and then sit there until the sun rose. After that we would clear the western boundary. Needless to say, none of us was very happy about the mission. None of us had gotten any sleep. But it was just another one of those things.

Dark Wing, Crowfoot, Nasty, and I took the south side of the CP while the rest of the guys took the north side. We were there to check out early morning traffic. Most of it was brick trucks moving stuff north. Some of it was the crazy stray dog herds that ran wild in our area. Packs ran everywhere and the ones we saw that night were big. Had the LT not been there, some of them would have gone to doggie heaven.

Nothing highlighted our night. I sat in the Hummer for the entire mission. Dark Wing and Crowfoot did do a few random car searches. We talked smack to one another but that was about it. When the sun came up we headed back towards town and then cleared the western route. After that we headed back to Caldwell. Things were going smoothly until we stopped to drop off Dark Wing. He didn't want to go to chow so I stopped the vehicle and let him out. We were getting ready to roll again when the Hummer wouldn't move forward. We heard a large grinding noise coming from underneath the hood. I tried again and the same thing happened - nothing.

I got out of the Hummer and looked under the engine and sure enough, one of the half-axel bars had fallen off. We couldn't go anywhere unless we put the vehicle in four-wheel drive. I had never experienced anything like it, and to be quite honest, I didn't know what in the hell to do next. We were able to move the vehicle to the side of the road where I waited for a tow vehicle to take it to the motor pool. Everyone else went to dump gear.

The mechanics hauled it over to our company's bay and downloaded it while I went to breakfast. Dark Wing had come back down and said he'd hang out until I got back. Breakfast was good and fast. I got back to the motor pool and saw that the mechanics were almost through with what needed to be done. One of the grease monkeys had said that if it wasn't too bad, it would only take about 20 minutes to fix. If it was bad, it could take all day. Fortunately for us, he was only under the vehicle until 0915.

What it boiled down to be was that screws and bolts had come loose on that particular piece of equipment. We had been told to check them every time we did preventative maintenance but apparently no one had done that lately on our vehicle. Murphy could have screwed us up really badly that morning but fortunately, the mishap happened when we were at a standstill on base. From what I was told, it could have really messed up the vehicle, maybe even us, had we been rolling balls to the wall down the road.

****

This particular section was written while I was pissed off and was meant for all the folks back home if anything had happened to any of us over the following two weeks. We were in the preparation stages for the elections. I thought that our Chain of Command, whether it was by directive of the 42nd ID or of its own initiative, had been making and continued to make stupid decisions that the soldiers who were under its authority would be paying for and had been paying for since the day we first stepped foot in this country.

Our command was going to put a company sized element in an unfortified position for five to six days during the elections. By unfortified, I mean no dug out positions, no coverings such as bunkers in case of mortar or RPG attack or some kid throwing a grenade over the top of the wall. Speaking of the wall, there was nothing there to stop some asshole from driving up in his VBIED and blowing it to hell. We kept being told of an immanent attack on the Youth Center that would happen while we stayed there. My question was why in the hell did we have to stay there in the first place then?

The whole idea of our being in the middle of Balad Ruz for that week, stuck in the Youth Center with that lack of security, just frustrated me. I knew that we had not lived like the troops had during the initial combat operations in Iraq. Hell, we lived better than most troops currently serving in the country. I thought the whole operation could have been planned and executed better. We had planned and executed operations from Caldwell for the past year and during past elections. Why did this election make any difference?

My time in Iraq made me reevaluate my desire to become an officer. I know it curbed my desire to learn the leadership skills that were taught to officers in Guard OCS. Very few men in my immediate chain of command impressed me as leaders. I still wanted to lead men but I was having a hard time deciding if I wanted to do it within the Guard. I continued to have a place for the military in my heart; my experience in Iraq could not diminish that, but I did not know how long I could have a positive attitude where I was.

October 8

At 2100 we were sent to check two polling sites that had questionable security. Of course, none of us wanted to do this because it was last minute and most of us were already transitioning into late night rest mode. But we had signed on the dotted line and were obligated to follow orders, so we geared up and rolled out.

We were given the mission brief by Rugger. This guy was an interesting character to say the least. I've already mentioned him but what I probably didn't say was that he one of those guys who in Vietnam would have more than likely been fragged by his own men for being a little too gung ho for his own good. He was too much of a "Yes, Sir" boy. I could understand being motivated to do one's best while serving; but there were times, especially as an officer, one had to make decisions without the chain of command's micro-management involved. He hadn't gotten to that point yet. We liked him but he did have his moments.

When he took over command of our platoon he made it a point to tell Dark Wing that he would not change much as far as our daily and routine operations but would want the guys to know that he was now in charge. From the instant he said that last word he changed everything. Of course, Dark Wing wasn't too happy with that word.

Rugger was like a kid in a toy store when he went out on patrol. He wanted to take everything he had at his disposal on every mission and wanted to use them just to say that he had. The funniest thing about his going on patrols was that he would wear a pair of catcher's shin guards that he'd brought. When he broke them out for his first patrol with us, we all about died. He looked like a dork and immediately we started riding him about them. He took our ribbing in stride but he kept on wearing those stupid shin guards.

We were slotted to ride to two polling sites and check their security. Rugger made sure our sniper team was fully stocked and took a full complement of personnel for a routine run. It was overkill. Ever since the Battle of Mogadishu and the movie *Black Hawk Down* U.S.

military leaders have been scared to death of getting their asses kicked by the general public of the nations in which troops operated. Rightfully so in some cases, but our guys weren't complacent when we rolled out. We knew what was expected of us.

Our missions were run with a minimum of three up-armored Hummers manned by at least three soldiers in each: driver, gunner, and vehicle commander. Sometimes there was an interpreter or medic riding as well. More often than not, we went out with three or four in our vehicle. We moved lightly so that we could move fast. The more equipment we took with us meant that we were that much slower. Sure we would have more firepower, but if we couldn't get to a place quickly enough to use it, we were dead in the water anyway.

We got to the first polling site at approximately 2300. The site was secured by nine Iraqi Police with AKs. In any other situation this would have been great security. Rugger radioed our TOC and let them know the situation and they gave us the order to go to the second polling site to check it. For the next two and a half, three hours, we were driving around the back roads and fields north of Balad Ruz, looking for that site. Our maps and GPS were not very helpful in the dark. We could not distinguish landmarks very well ahead of our turn time. Most of the roads that were on the map did not match with what was on the ground so we had to improvise some of our directions.

We finally got to the road that would take us to the second polling site and traveled down the east side of a shit canal that was very rough going. The route itself was full of runoff ditches and holes that had the potential to tear the undercarriage of a vehicle to shreds. Some of us dismounted and scouted the surrounding area for routes out and potential hazards along the way.

Rugger took four guys with him to see exactly where the building was. They approached on foot, and as any good security team would do with unexpected visitors, the Iraqi Police drew down on them but fortunately for everyone involved no one fired at anyone else. They finally realized that Americans were down on the ground and after an exchange of pleasantries, our guys walked away from the area with no other incident.

My interpreter & me smoking the Hooka

****

From: *<Malcolm>*
To: *<Becky><Dad><Mom>*
Sent: *Sunday, October 09, 2005*

Hey folks,

Just to let you know that I'll be out of pocket for the next few days because of our mission during the elections. Don't fret. I'll be back soon enough.

****

From: <Malcolm>
To: <Mom>
Sent: Thursday, October 13, 2005

Hey Marjorie … ha ha.

I got to come back to base for a couple of hours. Just wanted to let you and dad know things are fine. Stupid but fine. We'll be out for two more days and then all will be well. Combat operations for us are almost over. That's a bummer but at least we'll be out of the firing line for the most part. You never know until your feet are planted back in the States.

****

From: <Malcolm>
To: <Mom>
Sent: Saturday, October 15, 2005

Hey Mom,

Well, we're back from our little mission. It was long and a load of crap but hey, it's over. We could have done everything we did from the base but for some reason our commanders make the decisions and the rest of us have to live with them. Oh well.

Nothing big happened the whole time we were at the Youth Center in Balad Ruz for the elections. Today was pretty quiet. I'm glad. We had a patrol around midnight and then one at 0800 but nothing much was going on. Just checking the polling sites. Good security by the Iraqis. It was good to see them taking care of business.

Just a few more days and we're out of here.

****

From: *<Malcolm>*
To: *<Distribution List>*
Sent: *Sunday, October 16, 2005*

As I'm writing you, there are only six days left for our stay at Battle Position Caldwell, formerly known as Forward Operating Base Caldwell. Our last week was spent ensuring that the area we are charged with remained safe for voters to go to the polls and cast their vote on a referendum. I'm sorry to tell you this, but I had and continue to have no clue as to what they were voting for. For me, that's a good thing. It just keeps things simple – ensure the safety of the voters.

To be totally honest with you, when we were tasked to do this mission not many of us were excited about it. For almost a year we have been running patrols and missions in the town of Balad Ruz and its surrounding townships and countryside and have been very successful running them from Caldwell. This mission had us moved into the town for the week and staying at the Youth Center.

In most cases, staying at a youth center wouldn't be a bad thing, but in Iraq, we really have to watch out for our health and welfare. Because of the conditions of the city anyway, i.e., raw sewage and trash in the streets, over one hundred men living in close quarters adds that many more complications. Needless to say, we had a few guys who got hit with Saddam's Revenge by the second and third days. I'm still dealing with it – four days now!

There was no electricity when we got there so our mechanics hooked up some generators. They were great … when they worked. Not everyone was afforded the luxury of having electricity in the rooms in which they were staying. We had no air, no lights, no fans, and no working bathrooms. The port-a-johns were full by day four. Fortunately I had brought some citronella candles (can't remember who sent them to me, but thanks) and they kept the mosquitoes and flies away for the most part. The only consolation we had was in the form of a ping pong table. That definitely broke the monotony of trying to sleep in 100 degree heat in the rooms.

We ran missions day and night. Those were what I looked forward to. They gave me the chance to get away from "Fort Apache" plus

I got to see how things were going in preparation for the elections. In the two weeks leading up to the elections, some Shiite shrines had been bombed by the Sunnis. I hope I said that all right. Either way, one or the other bombed the others' shrines so we had to patrol the final one quite a few times. That got old going down the same route, but again, it got us out of the Fort.

This was what we did for the past week: patrolling the polling sites and making sure the people felt safe about the voting process. I think, overall, they did feel safe. The Iraqi Army and Police in our area did an outstanding job when crunch time came. For a few days we were worried, but they pulled up their boot straps and showed us their mettle. Maybe there is more hope than I give them credit for. I hope so. It's their country. It's their home.

Now we're just waiting to go home. I heard last night that another one of our brothers was killed by an IED, SGT Tucker [Robert Wesley Tucker of Hilham, TN]. I didn't know him but it bums me out all the same. From what I gathered, he was young, 20 maybe. My heart goes out to his guys and his family. This happened too close to the end (like anytime else would have been a better time). I also heard that three more guys in our platoon hit an IED a few days back. Fortunately all they had to deal with were stitches.

I'm so ready to come home. I got an email today from a buddy who got sent home a few months early because he blew his knee out. My heart broke reading his email about how he misses being over here and how he's still so proud that he IS a medic. Fillnutz told me that he'd been experiencing the emotions that come along with having gone to war, bonded with other soldiers in the trenches, and come home to find that regardless of how much people truly support you, unless they've been there and done it, they'll never really appreciate the "shit" that a soldier goes through. Pray for Fillnutz. Pray for all of us as we come back and try to re-acclimate to society.

So maybe there will be another update next month. I hope so. I'd like to think that life can always be an adventure like this. Well, not really like this, but you know what I mean. There are things that need to be done when I get home. I'll get to them in time. As for now, all I can think about is coming home and taking a vacation. Maybe the snow's falling in Colorado. I hope so. Good times. Thanks for letting me fill you in on what's happening here.

Again, one more week here, a few days transit time, and then we'll be in Mississippi. I'll see you soon.

<u>October 18</u>

We stayed in the Balad Ruz Youth Center for a week. It was lame. An entire company was in this building that had no running water, no working bathrooms, no showers, MREs, and minimal security. Our platoon was crammed into the workout room, twenty-five guys and all our gear, nut to butt. It was lame. As one would expect, we made the best we could of our situation.

Me, Little D, & Dark Wing relaxing at the Youth Center

Patrols were run twice a day. I always tried to go so I would not have to stay in the room. The weather still had temperatures in the high 90's during the day but it was better being on patrol than being stuck in that sauna back in Balad Ruz. We were able to make a few runs back to base during the week so we could take showers and grab a hot meal.

The only real issue that we had to deal with during the elections was the bombing of shrines for the few weeks preceding them. In Iraq, that was how one made his statement. He'd either kill somebody or blow another religious sect's shrine all to hell. It was asinine, I thought, but it was their culture and no amount of western influence was going to change their minds or hearts. God had to intervene and a reader of God's word would have known that the battle which was being fought in the Middle East had been waging for centuries. It was of Biblical proportion.

We did some crazy driving during our stay at the Youth Center. I suspect it was because we couldn't really relax there, so when we were out we did what we could to lighten the mood. On a few patrols we managed to get a Hummer or two stuck in mud or in some kind of ditch where we'd have to use the other two Hummers to pull the first one out. We had a lot of fun with it though. They were great vehicles to have in the desert. If I had the money and one was up for auction, I would buy it. Those vehicles proved themselves in my eyes. The high-profile ones back in the States are a joke.

I finally had the opportunity to ride with some of my mates from the COLT platoon. We did east route clearance. Monkey Wrench was in charge of the patrol. He cracked me up every time I saw him and was one of the first guys from the COLT Team who got a tattoo from me. He'd had a frog on a lily-pad that he wanted covered up with a piston. It turned out fine and he was happy with it. He and his guys would stop by often while I was doing ink. I was glad to get to know them. For the patrol, I rode in the truck with the K Train, Doo Doo, and DT.

We rode to Route V1 and checked out the new Iraqi Police control point. They appeared to have everything under control at the position, so we loaded back up and headed back to Caldwell.

October 19

We rolled out again this morning. Yesterday we were told that we would have a 0530 mission. At some point between the times I went to sleep last night and when I woke up this morning, the mission had been scrubbed. At 0600, Shrek came into the room and told us that the mission was on again. Dark Wing and I suited up and headed to the chow hall for a quick bite. We were to meet the guys there, load up, and then roll out.

The unit relieving us, the 101st Airborne out of Fort Campbell, had started arriving on post a few weeks prior. They weren't bad but there were so many of them that they made the lines to everything unbearable. They had sent a team of their senior NCOs to ride with us this morning. We loaded up the Hummers with our guys driving, gunning, and commanding while two 101st guys rode in the back of each vehicle. The idea behind this was to give them a "tour" of our area while they rode and took mental notes for when they took over. It gave them the opportunity to get out and see what was going on in the area and how we did things.

We headed east on Route T1, making a stop in the test fire pit. All weapons were hot and fired well. After the shoot, we pulled back on route and continued our mission. We still took for granted the security of the test fire pit. I wouldn't have been surprised if some jackass hadn't tried to place a bomb in there at some point. I would have. It was the perfect place because American convoys rolled right through it and I am sure they took the security for granted as well. My vehicle was in the lead and less than a click down the road we ran across some Iraqi Army vehicles on the side of the road. The soldiers were telling us that an IED had in fact gone off in the area where they had set up their cordon. They also showed us where one round of unexploded ordinance lay on the shoulder of the road.

We "sat on" the IED for about an hour and watched as first the IAs walked up to each of the sites and then the local populace drove over the explosion site as they began maneuvering to continue their route to wherever it was they were going. The IA never appeared concerned about checking out a possible IED. It was like they had no fear of the possibilities that one might blow. Maybe they weren't scared or maybe they simply didn't care if they lived or died. They would walk up to a possible bomb and check it out like it was just another object that would never go off. They were crazy. It was at about that time that some of us realized that there wasn't any real threat of another explosion. The IA left the area, giving us the thumbs-up for us to stay to hold off traffic and wait for EOD.

Now remember, we were less than a click away from base. We were told that EOD would be out in a couple of hours. It was unbelievable. Eventually I talked Rugger into allowing my gunner and me to drive down to the site and actually look at the UXO and IED hole firsthand. We took pictures of both sites for the reports that were sure to be written up about the situation. Nasty and I were in agreement that the UXO had actually been in the IED hole but was blown out when the explosion occurred. There was also a dead dog a few meters west of the hole. Old Sparky's days were up.

An eastbound convoy came our way while Nasty and I were checking the blast site. The lead vehicle stopped and we told the commander what was going on. I told him that we didn't think the UXO was a threat and that he could proceed but he would have to be the one who made that call for his convoy. We didn't want to hold them up for something that would later prove to be a lame duck. They agreed that it didn't appear to be a threat so they moved on through. It was a good call by their leader.

Nasty and I picked up some shrapnel from the blast and then headed back to the rest of the vehicles in our convoy. Rugger made the call to mark the UXO with a water box, which wasn't a bad idea, so that we could continue our patrol and leave the ordinance for EOD. But then he wanted to mark it with white smoke too. We had no idea why he wanted to do that. All we could come up with was that he just wanted to use up a smoke can. The 101st guys were laughing. I took some video of them as they carried out the LT's orders.

We loaded back up and headed west again to finish our route clearance. The rest of the mission was uneventful. When we came past the UXO/IED area, we noticed that EOD still hadn't gotten there to remove the ordinance. At least EOD was consistent - always hours behind schedule.

****

From: <Tates>
To: <Malcolm>
Sent: Wednesday, October 19, 2005

Hero work, Brother Mal. You and the boys have done marvelous things. Congratulations to you all – I'm proud of you, mate.

Hopefully by now you are eating your Pizza Butt pizza and Subway hoagies in Kuwait. If your convoy back is still pending, listen carefully, mate – be safe, but don't abandon what got you there and all the way safely to this point. Urgency to get across the berm can getcha killed. Speaking from experience – our recovery team, in making a routine tow hook up less than ten minutes into their final convoy from Anaconda back to Kuwait, accidentally ran over their platoon sergeant with a HEMTT wrecker. A mission they had done dozens of times and had done exceptionally well ended in absolute disaster because they were hustling too much. Imagine the blow to morale and to the crew ... it was devastating.

Be safe, but if you ever took the extra time to make sure every one is dialed in, this is that time and mission. Good luck, be vigilant, and be safe. Shoot us a note when you get a chance. This is the last task - make sure your boys don't get silly on the way out. You're in our thoughts and prayers, Bro.

OCTOBER 23

We just got released from another meeting. It was supposedly to put out information about our leaving Caldwell. It was information that we'd already been given. The questions that were asked of the First

Sergeant were answered by his saying, "I don't have that information now." He never did give a solid answer to anything during that meeting. It was a waste of time.

Brass also gave out border certificates to those soldiers who had been to the Iranian border at some point during their deployment. It was a lopsided event, mainly driven by Red Platoon and Headquarters. There were a few us who had been to the border but did not get our certificates. The piece of paper wasn't what I wanted. I wanted to be recognized by a chain of command that had finally made an effort to do things right by its soldiers. Once again, our administration and my chain of command had let me down.

I could tell that I was getting closer to the edge. I wanted to leave Iraq and the people whom I'd been working with for the past year. My attitude was beginning to reflect my animosity towards the whole situation. I needed to get away from everyone and everything for a while. Time was getting short. I just had to focus on that positive.

****

From: <Malcolm>
To: <Mom>
Sent: Monday, October 24, 2005

We should be flying out of Caldwell tonight. They, the ever present they, told us last night that we are supposed to be in the States by the morning of the 28th. We'll see. That's if everything goes perfectly. Never have seen that happen in the Army yet when it comes to movement plans.

I'll call you sometime that day so have your cells on or be around the phone at home.

October 26

We've been in route to get home for two days now. Two nights ago we left Caldwell. It was crazy. Hundreds of men lined up for hours waiting for choppers to come in and take us to Anaconda. We sat in a holding area from 1500 until after 2200. Tell me that wasn't crazy. We were able to keep ourselves entertained for the most part. Some of us threw rocks at targets to pass the time, the best being when another Joe went to the Port-a-Shitter. We would unload a barrage of rocks at the walls. I was certain that the guy inside was calculating if and when he should attempt to come out to return to his place in line. Some guys played cards, some slept, and others zoned out while listening to music. When the birds finally did come, we flew out two choppers at a time, 30 men per chalk.

Once at Anaconda, we got bussed to the transient area and were given billeting. After stowing our gear, most of us hopped a base bus for the PX area for some midnight chow at Burger King and Pizza Hut. I filled up on a Hawaiian Pizza and got the urge for a pizza run out of my system.

Joe on his way to the C14 that would take us out of Iraq

We returned to the shacks and crashed until our 0730 muster. After roll call some of us headed to the DFAC. Breakfast was good but cut short because we had to get back to the shacks to be ready to load up and head to the PAX terminal.

At 1000 we headed to the PAX terminal where we stayed until 2100 that night. We spent most of the day lounging around waiting for the flight time to arrive. A session of roasting took place half way through our stay when we found out that one of our boys, Nasty, had hooked up with an older female soldier on Caldwell in one of the Hajji marts. It was quite a funny story. Apparently little Nasty and another Joe had hooked up with a drunken female soldier as the Hajjis who ran the shop watched. The story itself wasn't as funny as watching him react to our trying to get the whole story out of him. We also found out that Man Love hooked up with a trans-gendered hooker in Thailand. We spent three hours riding both of them.

At 2100 we were told to grab our gear and head out to a roll call and then to the buses to be taken to the flight line. About 180 soldiers and civilians rode out on a C17. It was a good flight, only about an hour long. I had ridden one on the way back from leave. This time I rode on the sides in the jump seats. It wasn't a bad ride.

When we touched down in Kuwait we were dispatched onto some more buses and driven to Camp Victory. When I got off the bus I forgot my laptop in the overhead compartment. Fortunately, I caught the predicament rather quickly and the civilian female who was in charge of this particular part of our journey was able to call the bus driver and he gladly came back so that I could retrieve my computer.

Once at Camp Victory, we were offloaded and grabbed our gear. Next, we were corralled into another formation so that information could be put out as to what our schedule would be for the next 24 hours. After listening to 30 minutes of information we were directed to a holding tent where we hurried to grab a cot and wait for food to be brought to us since we had just missed mid-rats. We gorged on the scraps that our hosts had been able to round up and then hit the rack. Morning would come early considering it was after 0200 when we finally bedded down.

Home Bitter Sweet Home

November 1

We've been in the States for three days. It has been an emotional roller coaster. Some of it has been really good, elation beyond anything any of us have ever experienced and some of it has just been plain awful. We sat around with nothing to do on base except look at one another or go into town, if we had a ride, and spend our money on stupid things like clothes because the girl selling them was hot. Coming home has been everything and nothing.

Our trip from Kuwait took us through Shannon, Ireland again where the majority of us bellied up to the bar and tried to put down as many pints as we possibly could before the flight across the pond. It was a good time and true confirmation that our time in the Middle East was over. We wound down. There was nothing quite like looking out the window of the airplane and seeing Kuwait City and the Persian Gulf getting smaller and farther behind us. It was even better when we touched down on the Emerald Isle. Much of the tension that had been building up for the last month began to dissipate. The countenance of the soldiers was totally different than even 12 hours earlier.

Our second stop along the way was Bangor, Maine. It wasn't as powerful an experience as our stop in Shannon, Ireland, but it was good nonetheless. This was the first time our cell phones had service in a year. We were able to call friends and family with no jerk in the background yelling, "Number 7, you have five minutes! Number 2, your time's up!" Some of the people who worked the airport allowed some of the guys who didn't have phones to use theirs. We were able to snag a few more beers in Bangor.

Gulfport, Mississippi, wasn't quite the welcome I had expected. I actually hadn't expected anything, but we were greeted by the General of the Tennessee National Guard as well as by some family members and a small band. It was a nice gesture but unless we knew we were going to have friends or family down there, we just wanted to get out of there so we could crash. We were eager to begin our 48 hour stand down. The effort was appreciated by all soldiers, and we clapped after every song that the band played because we knew that they probably didn't want to be there anymore than we did.

When we got to Shelby, the first thing we did was turn in our weapons. While we were doing that a couple of the advanced party guys got the beer and whiskey ready for us at our barracks. You'd think we would have all just crashed because of such a long trip, but our second winds kicked in as we realized we did not have to be on alert anymore. It was a welcome relief and we stayed up until all the beer and whiskey were gone. We had a good time, telling each other how much we loved each other, how glad we were to be home, and how we'd stay in touch with one another for the rest of our lives. It was all psycho-babble BS except the part about being glad to be home.

I took the 48 hours of our stand down to go home and pick up my vehicle. Cow Tits' dad had left work in Tennessee when Cow Tits had called him from Maine. He showed up in Hattiesburg around five in the morning and five us loaded up in his car and headed back north. We dropped off Dark Wing and Pretty in north Alabama so that Dark Wing could spend some time with his dad. We dropped off another Joe on Lookout Mountain at his family's home.

When we pulled onto the road where Cow Tits' house was we could see that his two brothers and some of his friends were ready to assault him with love and admiration. It was a cool homecoming. I felt good for him. They seemed to be a tight group of rowdy brothers, the kind you'd enjoy getting into trouble with. We stayed at their house for a while as the family reunited and Cow Tits and his brother Greg revved up the Mustang that they had rebuilt. Cow Tits was on cloud nine.

We finally departed. Cow Tits and Greg drove me to Crossville to meet my parents who were patiently waiting for me at Cracker Barrel.

We said our hellos and goodbyes and then I jumped in the XTerra with Mom and Dad. We headed to Cookeville and arrived in time to say goodbye to the area high school kids who were going with Young Life to Sharp Top Cove for the weekend. I saw a lot of friends. It was good.

That night I went out for a while and ran into a friend at Spankies. Slim was there with some of her friends. We had a few drinks together and got caught up on life. A lot of folks I have known over the years were there as well. It was good seeing them and hearing how life had been treating them. Of course they asked me the same questions over and over: Are you glad to be back? How was Iraq? Do you think we really need to be there?

Saturday was spent on the couch watching college football and sleeping. It was good. I had plans to go out for a while but ended up only staying at my bud's house for an hour or two and then went back home to crash again. I guess the jet lag set in. No big deal though because Cookeville had much to be desired in the area of good times for a single male at 35.

Welcome Home!!!

November 11 Update

From: <Malcolm>
To: <Distribution List>
Sent: Friday, November 11, 2005

Well, folks, as you know, I'm home. It's good. It's been strange at times but good. Sometimes I still feel like I need to drive very aggressively to make my point on the road. Sometimes I miss having a gunner to clear vehicles out of my way as I drive along. Sometimes, and not too often, I miss my rifle.

I saw the movie Jarhead and one of the statements in the movie was about how a man can be trained with a weapon and have it with him for a year or so and no matter what he does after that, whether it be loving a woman or raising his kids, he'll always know how it feels to be one with a weapon. It's weird, yes, but a fact when you've gone day in and day out in a combat zone with a weapon on your hip and slung over your shoulder. I don't know how cops give it up when they retire after 20 or more years.

Sometimes I find I just like getting away from people ... kind of like I did in Iraq. But here I can travel so that gives me some good time to myself. I love seeing everyone that I can but sometimes, in certain situations, it's just better to be alone and not answer any questions. I think my mom and dad are worried sometimes, as if I might need to express myself more or show more emotion about what's going on. I don't know. That's just not me. Don't worry. It'll take more than Iraq to send me over the edge.

Sometimes I have a drink or two to sleep. I quit drinking a long time ago because I used to abuse the unholy hell out of it. That role John Belushi played in Animal House, yeah, that was I. So it was a good thing to quit for about ten years. But I've had a few since the enlistment. I don't ever want to get back to the way I was but it has helped a couple of nights just to fall asleep. Kind of like taking Ambian. I didn't sleep well in Iraq and very few times since being home have I felt like I got a good night's sleep. I think that's part of it though. Why would I sleep well in a combat zone? I had to be alert even when I was asleep if one can imagine that?

Sometimes I'm extremely happy. I got to spend the night at two friends' houses this week. One friend was a doctor who was in Iraq with us. He got to go home after a three month tour to keep his family practice going strong. He has a beautiful wife and six kids, one who was adopted from Guatemala. The other friend is an old fraternity brother and he and I had been at a reunion that night and then went back to his house to crash. He has a beautiful wife and two wonderful little kids. I woke up this morning saying hi to a beautiful little boy I had gotten the chance to talk to over the internet while I was in Iraq. That made me really happy.

I've also been able to talk to some great friends over the phone since I've been back. That's so good. I know life has gone at warp speed back home as I've been away and I can never jump in where I left off, but it sure is good to know they're still out there and still stoking the fire of friendship between us. That makes me happy.

I've been asked numerous times what I'm going to do now that I'm home. Am I going to go back to Young Life? Am I going to stay in the Guard? Am I going back to Iraq anytime soon? The most real answer I can give anyone is I don't really know at this point. I can see how some warriors of the past really got lost in life because of this. I'm not scared but if I really think about it for a long time, I have to be concerned about what's next. Without vision people will die. That's Scripture. I've already been anxious about having no mission after I get done with the leadership school I'm supposed to [attend] in December. In Iraq, I had a mission. Before Iraq, I had a mission. Now, I want to go skiing but that's about as far as my vision is cast at this point.

One of the things I do want to do is in the picture I'm attaching. I drew it up today. A soldier at what I call the "relaxed ready" in front of Old Glory - ragged and weathered - like I feel. I know God has gifted me with an ability to do art. I've done it off and on my whole life but I've never really pursued it with my whole heart. I want to do that. I don't know how it will happen. I do know that I can make a living doing tattoos but I want to do so much more. Dream big is what I've always been told. We'll see. I hope you all keep me in check on it.

Anyway, just wanted to let you all know what's been going on since coming home. Gator and the crew at the Country Giant, 94.7, you guys are amazing. Thank you for your endless love and support for the 278th soldiers and their families. Families don't get enough recognition for the sacrifices they make to be away from their soldiers, Marines, Sailors, or Airmen. Thank you all.

I learned a lot of things about myself while I was in Iraq. I realized the hard way that I wasn't as perfect as I once thought. I found that I struggled quite a bit when interacting with others who did not live up to the expectations I had of them especially if those persons appeared to have no expectations of themselves. I struggled with some of the leadership decisions that were made involving the lives of men who had no part in making those decisions. I don't think that was abnormal for any of us to feel that way.

In hindsight, I'm able to look back and say that nothing was perfect while we were in Iraq. I was crazy to expect that. Nothing would ever be perfect. If it had been, we wouldn't have been there in the first place. Some things were just more frustrating than others. Some situations I handled well, others I didn't.

As for being prepared to go to war, I felt like the Army did a hell of a good job on me at Fort Sam Houston. The medical training was second to none. Nothing I saw in theater was new to me. The cadre had trained and conditioned me well for what I would experience in combat. The medical training was and continues to be the foundation that is helping to save the lives of so many warriors out on the world's battlefields.

The biggest thing that has impressed me has been the resiliency of America's troops, in our particular case, Tennessee National Guard personnel. Our men and women wanted to serve and then get back home. Our goal wasn't just to go to Iraq to help rebuild a nation. Our goal was similar to that of extreme mountaineers. Our goal was to go there and return to base camp in one piece. That's the goal of every soldier. Whether we were actually fighting a war on terrorism like so many of us had joined for or we were in fact helping Iraq to rise up like the Phoenix, out of its ashes to fly high again as a great nation, we all wanted to make it home.

The majority of the 278th made it home alive. Some of our brothers didn't. A few of our brothers will physically deal with the war for the remainder of their lives. We'll all deal with it to some degree psychologically. That's part of it. We knew we all ran the risk before we left the house. Now we need to allow time, patience, and love to heal those wounds. Love is the greatest of these.

Leadership Lessons Learned

1. Never allow persons to remain in a position who are not qualified to fill that position and by that I mean don't allow them to stay in the position because they have rank, seniority, or you're just too damn lazy to shake things up. If they are not performing the duties required to accomplish the mission of that position, then it's your responsibility as a leader to take them out of that position and replace them with someone more competent. If no one is more competent, then it falls on your shoulders to assume that position until you can find someone to fill the vacancy. It's called responsibility.
2. Discipline. Make the punishment fit the crime. Ensure that discipline is carried out quickly and justly. Inform all those involved in the discipline process why the individual is being disciplined. And as far as mass punishment, that's poor leadership. It tells your subordinates that you have no real control of their actions nor do you want to hold an individual accountable for his or her actions. By not holding individuals accountable for their actions, you are teaching them to make excuses, place blame on others, and not take personal responsibility.
3. Discipline. It's not always a bad thing. Which pain do you live with – discipline or regret? Without discipline nothing that has been accomplished would ever have been accomplished. Discipline is the key to success in life. In everything we do in life some form of discipline shapes our approach to that function or goal.

4. A senior leader should not base his actions on the agenda of the mob, those who are junior to you. By that I mean, if you're just trying to be everyone's buddy, just so you can feel like you have a close knit group of "friends," rather than being a leader, then doom on you. All the other personnel who are doing what's necessary to make the machine run smoothly under your command will pick up on that and respond accordingly by not supporting you. You will lose positive energy for the sake of a few bad apples. It's your job either to train those bad apples or eliminate them from the equation if they are not becoming part of the solution. It's not your job to baby them. They have a mom for that.
5. A leader should not spend all his or her time working and dealing with the bad apples. If they're being assholes, discipline them and move on because when you spend all your time and energy with the bad apples, your good apples quickly lose interest in the team, supporting you, and the mission. Spend as much time or more praising the accomplishments of those who have done good work and stayed out of trouble as you do with those who make no effort to elevate the team.
6. Take care of your men. Don't sacrifice their well-being simply for the sake of your own glory. Too many times leaders are overly concerned about climbing the corporate ladder at the expense of those who've helped them get there. Just don't do it. You might end up with a grenade in your lap.
7. Ensure that your subordinates understand the bottom line of their mission clearly. It's up to every leader in the company to make every effort to keep his people in the know about what's going on.
8. Attitude is everything. If you want a bunch of dirt bags, just be a dirt bag yourself. If you want a bunch of go getters, be the best leader you can be – be a go getter. Always lead from the front and always lead by example because regardless of

whether or not you think they are, your people are watching you and will do what you do more often than what you say. What you do in moderation, they will do in excess.

9. With some exceptions, your people are competent if not intelligent. They would not be in the position they are in if they were not. Give them the benefit of the doubt until they prove you wrong. Treat them as competent human beings. Don't have them doing stupid stuff simply for the sake of making them look busy. There are stupid questions and there are stupid assignments. Keep them both to a minimum.
10. Encourage your subordinates to strive for every goal they can attain in a given situation. Find ways to make your people better and reward them for their efforts. If they achieve their goal, that will be their reward; but if they don't achieve the goal through honest effort, at least reward their efforts so they will be encouraged to continue to strive for that goal and others in the future.

Glossary

AIF

- Anti-Iraqi Forces (Ali Baba – the fanatics who senselessly kill other Iraqis and attempt to disrupt the efforts of Coalition Forces from helping the Iraqi people, all in the name of Allah)

AK

- From Wikipedia, the free encyclopedia - The AK-47 (shortened from Russian: Автомат Калашникова образца 1947 года, *Avtomat Kalashnikova 1947*) is a gas-operated assault rifle designed by Mikhail Kalashnikov, and produced by Russian manufacturer Izhevsk Mechanical Works and used in many Eastern bloc nations during the Cold War. It was adopted and standardized in 1947. Compared with the auto-loading rifles used in World War II, the AK-47 was generally more compact, with a shorter range, a smaller 7.62 × 39 mm cartridge, and was capable of selective fire. It was one of the first true assault rifles and remains the most widely used and known. More AK-47 rifles and variants have been produced than any other assault rifle; production continues to this day.

AO

- Area of Operation – Any place where we conducted operations; sometimes used as slang in conversation, "I'm in your AO" meaning "I'm in your immediate vicinity."

ARCOMM

- From Wikipedia, the free encyclopedia- {The Army Commendation Medal} The Commendation Medal is a mid-level United States military award which is presented for sustained acts of heroism or meritorious service. For actions where such performance was in direct contact with an enemy force, the Valor device ("V" device) may be authorized as an attachment to the decoration.

BUD/S

- Basic Underwater Demolition Seal Training - The first phase of United States Navy Seal training held in Coronado, California. Seals are the Navy's elite commando force.

CFB

- Combat Fitness Badge – The CFB is not an actual military award but a nickname given to medic who I had become good friends with at Fort Sam Houston during our medical training.

CHU

- CONEX Housing Unit – These were the little shipping containers turned housing units which most of us lived in throughout our tour in Iraq. The CHU could hold four soldiers but we were fortunate in that we only had to share our "homes" with one other soldier.

CLS

- Combat Life Saver – These are the soldiers in the platoon who have either volunteered or been selected by their leadership to get basic first aid and trauma care training so that, should an incident occur, they can assist the medic in the platoon.

CO

- Commanding Officer – Usually the CO is at the Company level and higher: Company Commander, Squadron Commander, Regimental Commander, etc.

COLT

- Combat Observation and Lasing Team

CP

- Check Point – Also known as a Tactical Check Point, these were designated positions which served as check-in points so that if communications went sour, the Tactical Operations Center would know a patrol's approximate position.

CRMC

- Cookeville Regional Medical Center – The men and women of this facility supported our local soldiers with letters and packages.

DCU

- Desert Camouflage Uniform

DFAC

- Dining Facility – The place where we ate chow; prepared and served four times each day by contractors.

EOD

- Explosive Ordinance Disposal – Elite teams of bomb experts. They were called anytime ordinance was found along any of our patrol routes.

FLA

- Fleet Light Ambulance – Converted Hummers used to transport casualties to and from aid stations.

FOB

- Forward Operation Base – any number of bases throughout theater where troops were stationed and conducted military operations.

HEMTT

- Heavy Expanded Mobility Tactical Truck

HHT

- Headquarters / Headquarters Troop – The main administrative and operations unit of the Squadron.

HQ

- Headquarters – The central location where operations were conducted from and information collected and disseminated.

HVAC

- Heating, Ventilating and Air Conditioning

IA

- Iraqi Army – The soldiers of Iraq who we were there to train and establish as the military force in that country.

IBA

- Individual Ballistic Armor – These were the vests we wore while in country. Many lives have been saved because of them but they have also given our country a high number of extremity casualties.

IED

- Improvised Explosive Device – The main weapon used by Anti-Iraqi forces. They could be made of anything and could be placed anywhere.

IM

- Instant Messaging – Internet real-time messaging. It was better than email in that if you were lucky and were able to be on a computer with someone back home, you could have a conversation with them rather than simply sending an email and waiting for a reply.

ING

- Iraqi National Guard – We worked with these soldiers the most. More often than not, they were the best ones to work with because they were just like us, Guard.

IP

- Iraqi Police – The most corrupt of the Iraqi forces. Very seldom did we trust these guys.

JCC

- Joint Communications Center (also referred to as the JOC – Joint Operations Command) – This was a relay center for long range patrols and was a center which we guarded the duration of our tour.

KBR

- KBR (formerly Kellogg Brown and Root) is an American engineering and construction company, formerly a subsidiary of Halliburton, based in Houston, TX.

KIA

- Killed In Action

LMTV

- Light Medium Tactical Vehicles

LSA

- Logistics Support Area

MiTT

- Military Transition Teams – Soldiers and Marines embedded in the Iraqi Army to assist in the training and the evaluation of progress made by the Iraqi soldiers.

MOS

- Military Occupational Specialty – The job indicator for every soldier, Marine, airman, or seaman, i.e., 91W is a field medic.

MOUT

- Military Operations Urban Terrain – Training held inside a simulated urban setting.

MRE

- Meal Ready to Eat – The backbone of the military's fine dining; extended storage capable and are usually eaten on extended patrols or anytime the military does not make contractors available at a given base.

NCO

- Non-Commissioned Officer – The backbone of the military. These men and women are the leaders of the enlisted ranks and ensure that all types of military operations and orders are carried out.

NCOIC

- Non-Commissioned Officer in Charge – The ranking NCO of any operation or evolution. The NCO is responsible for the men and women under him or her and all actions taken to accomplish the mission.

NIA

- New Iraqi Army

NTC

- National Training Center – Fort Irwin, California

NVG

- Night Vision Goggles – This technology allows military operations to be accomplished more effectively at night.

OCD

- Obsessive / Compulsive Disorder

OCS

- Officer Candidate School

ODA

- Operational Detachment Alpha – The first fire team of a Special Forces unit in country.

OPSEC

- Operational Security – The means by which all operations are kept on a "need to know" basis.

PAX

- Any number of personnel on a mission. The term is most often used for head counts.

PX

- Post Exchange – The place where we purchased toiletries and other comfort items.

RCO

- Regimental Commanding Officer

RIP

- Relief In Place – The time when two units in transition overlap their transition time. The departing unit gives the arriving unit "hands on training" in their Area of Operation.

ROE

- Rules of Engagement – The rules every American must follow in the process of engaging an enemy or known threat.

RPG

- Rocket Propelled Grenade

SCO

- Squadron Commanding Officer

SF

- Special Forces

SOP

- Standard Operating Procedures

SP

- Start Point

TC

- Track Commander / Tank Commander – The senior ranking NCO or Officer in a vehicle.

TCP

- Tactical Check Point

TMC

- Troop Medical Center

TOC

- Troop Operations Center / Tactical Operations Center

USO

- United Service Organizations – The USO is a private, nonprofit organization whose mission is to provide morale, welfare and recreation-type services to our men and women in uniform. Its "style of delivery" is to represent the American people by extending a touch of home to the military. The USO currently operates more than 130 centers worldwide. (from www.uso.org)

UXO

- Unexploded Ordinance

VBIED

- Vehicle-borne Improvised Explosive Device

XO

- Executive Officer – The second in command under the commanding officer.

About the Author

Malcolm Rios has served with the 278th Regimental Combat Unit for three years, one year of which was served in Iraq. He has previously served in the United States Navy. With over 10 years of combined military experience, Rios is able to articulate the ups and downs of the common "Joe" in a combat environment, all the while, providing the reader with an idea of what the situation on the ground in Iraq was really like for some of those serving in National Guard units.

Rios received his undergraduate degree from Tennessee Technological University and his post-grad degree from the Assemblies of God Theological Seminary.

Printed in the United States
79128LV00004B/1-87

9 781425 928698